DEVELOPMENTS IN FOOD ANALYSIS TECHNIQUES—3

CONTENTS OF VOLUMES 1 AND 2

Volume 1

Volume 2

DEVELOPMENTS IN FOOD ANALYSIS TECHNIQUES—3

Edited by

R. D. KING

*National College of Food Technology
(Department of Food Technology),
University of Reading, UK*

ELSEVIER APPLIED SCIENCE PUBLISHERS
LONDON and NEW YORK

ELSEVIER APPLIED SCIENCE PUBLISHERS LTD
Ripple Road, Barking, Essex, England

Sole Distributor in the USA and Canada
ELSEVIER SCIENCE PUBLISHING CO., INC.
52 Vanderbilt Avenue, New York, NY 10017, USA

British Library Cataloguing in Publication Data

Developments in food analysis techniques.—3
1. Food—Analysis—Periodicals
664′.07 TX545

ISBN 0-85334-262-8

WITH 20 TABLES AND 42 ILLUSTRATIONS

The selection and presentation of material and the opinions expressed in this publication are the sole responsibility of the authors concerned.

Printed in Great Britain by Galliard (Printers) Ltd, Great Yarmouth

PREFACE

The complexity of modern analytical methods is such that considerable laboratory experience is required in order to obtain a working knowledge of one particular area of analysis. The time spent gaining expertise can be reduced if we can benefit from the experience of the specialist. I hope that this volume will help to promulgate some of the insight the authors have in their area of specialism.

In this volume, Selvendran and DuPont consider the problem of estimating dietary fibre in food. Many laboratories are concerned with developing new methods for determining fibre, but exactly what constituents of foods should be considered to be fibre? In the following chapter some aspects of trace metal analysis are considered by Wolf and Harnly. Levels of essential as well as toxic metals are very much of interest, especially as the same metal can be beneficial at low levels and toxic at higher levels. Sampling and sample preparation pose particular problems in the detection and estimation of mycotoxins in food. Van Egmond reviews these problems before examining in detail the various approaches to the determination of mycotoxins. Another group of contaminants frequently found in food are the pesticides. A food-related approach to their analysis is described in the chapter by Young. The final chapter by Daussant and Bureau looks at the potential of immunological methods for food analysis. The progress in monoclonal antibody production is making this methodology more accessible to the routine laboratory.

R. D. KING

CONTENTS

LIST OF CONTRIBUTORS

DANIELLE BUREAU

Centre National de la Recherche Scientifique, Laboratoire de Physiologie des Organes Végétaux, 4 ter route des Gardes, 92190 Meudon, France

JEAN DAUSSANT

Centre National de la Recherche Scientifique, Laboratoire de Physiologie des Organes Végétaux, 4 ter route des Gardes, 92190 Meudon, France

M. SUSAN DUPONT

Agricultural and Food Research Council, Food Research Institute, Colney Lane, Norwich NR4 7UA, UK

H. P. VAN EGMOND

Laboratory for Chemical Analysis of Foodstuffs, National Institute of Public Health, PO Box 1, 3720 BA Bilthoven, The Netherlands

JAMES M. HARNLY

Nutrient Composition Laboratory, United States Department of Agriculture, Human Nutrition Research Center, Beltsville, Maryland 20705, USA

ROBERT R. SELVENDRAN

Agricultural and Food Research Council, Food Research Institute, Colney Lane, Norwich NR4 7UA, UK

WAYNE R. WOLF

Nutrient Composition Laboratory, United States Department of Agriculture, Human Nutrition Research Center, Beltsville, Maryland 20705, USA

R. W. YOUNG

Department of Biochemistry and Nutrition, College of Agriculture and Life Sciences, Virginia Polytechnic Institute and State University, Blacksburg, Virginia 24061, USA

Chapter 1

PROBLEMS ASSOCIATED WITH THE ANALYSIS OF DIETARY FIBRE AND SOME RECENT DEVELOPMENTS

ROBERT R. SELVENDRAN and M. SUSAN DUPONT

AFRC Food Research Institute, Norwich, UK

1. INTRODUCTION

1.1. Definition of Dietary Fibre

Dietary fibre (DF) was defined by Trowell[1] in 1972 as 'that part of plant material in our diet which is resistant to digestion by secretions of the human digestive tract'. As this definition did not include polysaccharides present in some food additives such as gums, algal polysaccharides, pectins, modified celluloses and modified starches, Trowell *et al.*[2] extended the definition to include 'all the polysaccharides and lignin that are undigested by endogenous secretions of the human digestive tract'. Accordingly, for analytical purposes, the term DF refers mainly to non-starchy polysaccharides and lignin in the diet. Other non-digestible substances present in relatively small amounts in plant cell walls, but nevertheless important in determining their overall properties, which may be included in a broader definition of the term DF are cell wall (glyco)proteins, cutin, waxes, phenolic esters and inorganic constituents. Although DF constituents, as defined above, may survive digestion in the mouth, stomach and small intestine, some of the constituents may be degraded by micro-organisms of the human colon.[3,4]

1.2. Occurrence, Distribution, Structure and Chemistry

A knowledge of the occurrence, distribution, structure and chemistry of the main DF polymers helps us to appreciate (1) the experimental problems involved in their analysis and how these may be overcome, and (2) the

TABLE 1
COMPONENTS OF DIETARY FIBRE IN A MIXED DIET

Main components of a mixed diet	*Tissue types*	*Main constituent groups of DF polymers*
Fruit and vegetables	Mainly parenchymatous	Cellulose, pectic substances (e.g. arabinans, rhamnogalacturonans), hemicelluloses (e.g. xyloglucans) and some glycoproteins
	with some lignified tissues	Cellulose, lignin, hemicelluloses (e.g. glucuronoxylans) and some glycoproteins
Cereals and cereal products	Parenchymatous and	Hemicelluloses (mainly arabinoxylans and in some tissues, e.g. barley, β-D-glucans), cellulose, phenolic esters and (glyco?)proteins
	lignified tissues	Hemicelluloses (mainly glucurono- or 4-*O*-Me-glucuronoarabinoxylans), cellulose, lignin and associated phenolic esters and (glyco?)proteins
Seeds other than cereals (e.g. leguminous seeds)	Parenchymatous (e.g. cotyledons of peas) and	Cellulose, hemicelluloses, pectic substances and glycoproteins
	some cells with thickened endosperm walls (e.g. guar)	Galactomannans, and small amounts of cellulose, pectic substances and glycoproteins
Food additives		Gums: guar gum, gum arabic, alginates, carrageenans, xanthan gums, modified celluloses, modified starches, etc.

requirements imposed by the diversity of starting materials. Most of the methods have important limitations, which must be considered in the interpretation of analytical results.

The main components of DF in a mixed diet are summarised in Table 1; this lists the types of polymer that can be obtained from parenchymatous and lignified tissues of fruits, vegetables, cereal products and seeds (other than cereals) and also some food additives. The parenchymatous tissues can be regarded as having (mainly) primary cell walls, whereas lignified tissues have cell walls which have ceased to grow and have undergone secondary thickening. Secondary walls are, in general, considerably thicker than primary cell walls. The main cell wall polymers of parenchymatous tissues are pectic substances, hemicelluloses and cellulose, whereas those of lignified tissues are lignin, hemicelluloses and cellulose; the types of hemicellulosic polysaccharides present in the cell walls of both types of tissue are usually different. So, while pectic substances predominate in the cell walls of parenchymatous tissues, lignin is a major component of the cell walls of lignified tissues.

In terms of structure, the constituents of cell walls fall into three groups: the fibrillar polysaccharides, the matrix polysaccharides and the encrusting substances.[5,6] Fibrillar and matrix polysaccharides are formed simultaneously during the wall formation, whereas the encrusting substances, namely lignins, are formed during secondary thickening of specialised cells. The fibrillar polysaccharides, which are the basic structural units of microfibrils, are a remarkably constant feature of the cell walls of all green plants, and in nearly all species these microfibrils are made up of cellulose. The α-cellulose fraction isolated from most plant tissues, however, usually contains small but significant amounts of non-glucan polysaccharides and glycoproteins associated with it. This association of cellulose with other polysaccharides might either occur during the preparation of the material by some adsorption effects, or it might form part of the essential organisation of the polysaccharides within the cell wall. If it is the latter, then the associated polysaccharides probably serve as keying material for the entanglement of the cellulose microfibrils with the matrix polysaccharides.

The matrix polysaccharides are made up of linearly oriented polymers which are present at all stages of the development of the wall and also of highly branched polysaccharides that are deposited at particular stages of growth. These polysaccharides may, at the surface of the microfibril, be incorporated into its structure. There are two major fractions in the matrix polysaccharides: (1) the pectic substances, which are soluble in water or solutions of chelating agents for Ca^{2+} ions; and (2) the hemicelluloses, which represent the polysaccharide fraction soluble in alkali. Differential extraction procedures are convenient, but not entirely selective, and the classification of polysaccharides is best based on structure rather than mode of isolation. The relationship of the operationally defined groups of polysaccharides to structural families, and the main sugars and glycosidic linkages present in some of the DF polysaccharides, are shown in Table 2. For an account of the chemistry of cell wall and DF polysaccharides see refs. 7–9. In addition to polysaccharides, plant cell walls contain glycoproteins, both hydroxyproline-rich and hydroxyproline-poor, and the associated sugars are mainly arabinose and galactose. Cell wall proteins could account for about 3–8 % of the weight of the cell wall material of most types of tissue.[10–12]

Dietary fibre polysaccharides may therefore be defined and classified in terms of the following properties: (a) kinds of monosaccharides present; (b) monosaccharide ring forms (furanose or pyranose); (c) positions ($1 \rightarrow 2$, $1 \rightarrow 3$, $1 \rightarrow 4$, $1 \rightarrow 6$) of the glycosidic links; (d) configuration (α or β) of the glycosidic links; and (e) degree of linearity or branching of the chains. Other relevant properties are molecular size, shape, crystallinity, solubility in various solvents and susceptibility to acid hydrolysis. The degree of branching varies from none, through a few single monosaccharides as side units along a main chain, to complex multiple-branched polymers. The polysaccharides may be completely homogeneous or highly heterogeneous and may be crystalline or amorphous. The monosaccharides commonly present in cell wall polysaccharides are D-glucose, D-galactose, D-mannose, L-arabinose, D-xylose, D-galacturonic acid, methylesterified D-galacturonic acid, D-glucuronic acid and its 4-*O*-methyl ether, and smaller amounts of L-rhamnose and L-fucose. These monosaccharides are all in the pyranose form except for L-arabinose which is commonly furanose; β-glycosidic links dominate with $\beta(1 \rightarrow 4)$ links probably the most common.

Among the non-saccharide components of cell walls, lignin stands out as a unique aromatic polymer. Lignin is probably deposited on cellulose microfibrils. Lignification commences in the region of the middle lamella and then spreads gradually towards the cell membrane. Lignin occurs to the extent of 15–35 % in most supporting tissues of higher plants and is therefore a major component of the xylem, the sclerenchyma of phloem and certain specialised cells, e.g. parchment layers of runner bean pods. Lignin is probably covalently linked to at least some of the hemicelluloses and is commonly recognised by the red coloration which it develops on treatment with phloroglucinol followed by HCl.[13,14]

Work on the chemistry of lignin indicates that it is a polymer consisting for the most part of substituted phenylpropane (C_6C_3) units: 3-methoxy-4-hydroxyphenylpropane (guaiacyl), 3,5-dimethoxy-4-hydroxyphenylpropane (syringylpropane) and 4-hydroxyphenylpropane (*p*-coumaryl) residues. Plant lignins of interest can be divided into three broad classes, which are commonly called softwood (gymnosperm), hardwood (dicotyledonous angiosperm) and grass (monocotyledonous angiosperm) lignins.[13,14] Of these the softwood lignins contain mainly guaiacyl residues, the hardwood lignins contain syringylpropane residues in addition to guaiacyl residues, and the monocotyledonous angiosperm lignins contain syringylpropane and *p*-coumaryl residues in addition to the guaiacyl residues. The angiosperm lignins, which are the ones of interest in the dietary fibre context, demonstrate considerable variation from species to species:

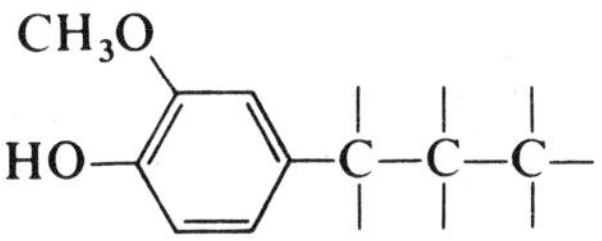

Guaiacyl

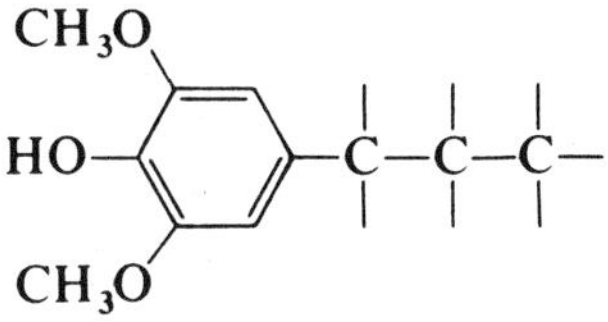

Syringylpropane

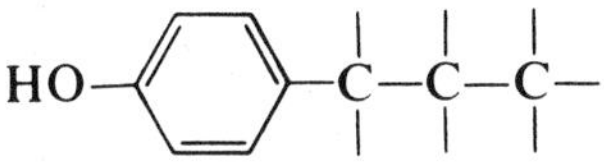

p-Coumaryl

1.3. Scope of the Review

This chapter is not intended to be a comprehensive record of methodology for DF analysis, but outlines the methods that have been used and discusses some of the problems that may be encountered and how they may be overcome. Such information would help food researchers and analysts to assess the relative merits of the published methods and the factors that have to be borne in mind when choosing, or devising, the most appropriate method for a particular food product. Since most food products contain a range of DF polymers, the complete analysis of the components would be a major analytical undertaking. This problem is compounded for mixed diets. Hence, any analytical procedure for DF must represent a compromise between the ideal, that is complete separation and analysis of all the components (which is very difficult and time-consuming), and a more empirical, simplified approach, which gives an estimate and an indication of the main groups of DF polymers present. Because DF analysis, as we understand it today, deals mainly with the latter type of methodology, we shall discuss it in greater detail. There is good reason to hope, however, that within a few years the problems associated with the simplified methods of analysis of DF will have been solved. The emphasis may then well shift to the problem of defining the constituent polymers, their physicochemical homogeneity, the specific properties of the polysaccharides (and associated compounds), and their numerous effects on the activities of the human

TABLE 2

STRUCTURAL FEATURES OF SOME OF THE MAJOR DIETARY FIBRE POLYSACCHARIDES

Polysaccharide	*Monosaccharides*		*Structural features*
	Major	*Minor*	
Cell walls:			
Pectic acid (pectin)	D-Gal A	L-Rha, L-Ara, D-Gal	α(1 → 4)-linked D-Gal pA (or methylesterified Gal pA) residues containing (1 → 2)-linked β-L-Rha p; galactan and arabinan side-chains
Xyloglucans	D-Glc	D-Xyl, D-Gal, L-Ara, L-Fuc	β(1 → 4)-linked D-Glc p residues containing α-D-Xyl p residues as side-chains on C-6 of at least half the Glc residues, with some of the Xyl residues being further substituted with β-D-Gal p, α-L-Ara f and L-Fuc p residues
Glucuronoxylans (and 4-*O*-Me-glucuronoxylans)	D-Xyl	D-GpA, 4-*O*-Me-D-GpA	β(1 → 4)-linked D-Xyl p residues with side-chains of D-GpA or 4-*O*-Me-D-GpA
Arabinoxylans (and glucuronoarabinoxylans)	D-Xyl	L-Ara, D-GpA, 4-*O*-Me-D-GpA	β(1 → 4)-linked D-Xyl p residues, with side-chains containing (1 → 3)-linked L-Ara f and smaller amounts of (1 → 2)-linked α-D-GpA or 4-*O*-Me-D-GpA residues

Galactomannans	D-Man	D-Gal	β(1 → 4)-linked D-Man p residues, with single unit side-chains containing α-D-Gal p residues
β-D-Glucans	D-Glc		β(1 → 4) and β(1 → 3)-linked D-Glc p residues
Cellulose	D-Glc		Linear chains of β(1 → 4)-linked D-Glc p residues
Food additives:			
κ-Carrageenan	3,6-Anhydro-α-D-Gal p, β-D-Gal p sulphated at *O*-4		Consists of alternating sequence of *O*-4-substituted 3,6-anhydro-α-D-Gal p and *O*-3-substituted β-D-Gal p which is sulphated at *O*-4
Alginate	D-Mannuronic acid L-Guluronic acid		Consists of β-D(1 → 4)-linked mannuronic acid and α-L(1 → 4)-linked guluronic acid
Xanthan gum	D-Glc, D-Man, D-GpA		Has a backbone of β(1 → 4)-linked D-Glc p residues; the side-chains are composed of two D-Man p residues and one D-GpA residue and have the structure: β-D-Man p-(1 → 4)–β-D-GpA-(1 → 2)–α-D-Man p-(1 → 3)–backbone; the ratio of D-Glc: D-Man: D-GpA = 2:2:1 (GpA, glucuronic acid)

digestive tract. Therefore some aspects of the methods which throw light on the finer details of DF structure will also be given. These will include improved methods for preparing and analysing gram quantities of DF from both fresh and processed foods. For both types of methodology, we have used recent developments in our laboratory to illustrate certain points.

2. PROBLEMS ASSOCIATED WITH THE ANALYSIS OF DIETARY FIBRE

The problems associated with the analysis of DF are discussed under three headings: (1) those arising from the preparation of DF; (2) those associated with the determination of the constituent carbohydrates; and (3) those encountered in lignin determination.

2.1. Preparation of Dietary Fibre from Fresh and Processed Foods

The preparation of DF from storage organs of edible plant tissues, which are usually rich in starch and proteins, is difficult. In oats and potatoes, for example, the starch to proteins to DF ratios are about 7·7:1·6:1·0 and 15:0·7:1, respectively.[15,16] Therefore incomplete removal of starch would seriously affect the values for DF. With processed starch-rich products there are additional problems arising from modifications of the starch, interactions between starch, proteins, DF and other intracellular compounds, the net effect of which is to render the starch less degradable enzymatically. Of particular significance are retrograded starch and starch–protein–lipid complexes. Under certain conditions of heat treatment (by extrusion cooking, etc.), water is eliminated between adjacent OH groups of starch molecules and the resulting 'starch complex' is irreversibly modified and may not be completely degraded enzymatically. The presence of modified starches, in the partially purified DF preparations, makes meaningful estimation of the β-glucan content of cereal products difficult. Further, the modified starches contribute to the total DF content of the product. Incompletely removed proteins interfere with lignin determination. The problem is further complicated with polyphenol-rich tissues, because the proteins bind polyphenols and their oxidation products, thus making meaningful estimation of lignin difficult.

The starting materials for the preparation of dietary fibre are usually alcohol-insoluble residue (AIR),[16–19] hot neutral detergent residue (NDF),[20–22] cold neutral detergent residue[23] and freeze-dried material.[24]

Although specific comments on the various methods using the above preparations will be made in Section 3, some general comments will be made at this stage. The alcohol-insoluble residue (AIR) is easy to prepare and handle, and is useful for fresh (and processed) foods which are relatively poor in starch and proteins, such as leafy vegetables and most fruits. For starch- and protein-rich products, complications may arise because of co-precipitation effects. The reasons for this are twofold: (1) The starch released from the granules is insoluble in the solvent and undergoes a retrogradation type of effect. This effect arises mainly because of the strong tendency for H-bond formation between OH groups on adjacent starch molecules, but some interaction with cell wall polymers and intracellular proteins may also occur. Some of the co-precipitated starch may then become resistant to enzymatic degradation and is not readily solubilised by aq. DMSO (dimethylsulphoxide).[16] (2) The alcohol treatment causes strong interactions (possibly via H-bonding effects) between the intracellular proteins and cell wall polysaccharides, and thus alters their properties. The co-precipitated proteins are not readily solubilised by pronase, or by aqueous organic solvents, e.g. phenol/acetic acid/water and aqueous detergents.

In addition to co-precipitation effects, incomplete disruption of the tissues is also a factor which hinders complete removal of starch and proteins.[15,16,23,25] Light microscope studies have shown that blending a plant tissue in an Ultraturrax and wet ball-milling the triturated material are necessary to disintegrate the cell structure and render the contents accessible to enzymes and solvents. This point has been illustrated with suitable examples elsewhere.[15,23]

Neutral detergent solutions (1·5 % w/v Na lauryl sulphate or 1 % w/v Na deoxycholate), hot or cold, are good solvents for intracellular proteins. The hot neutral detergent residue suffers specifically from both the loss of an indefinite amount of pectic substances, β-glucans and arabinoxylans, and the incomplete removal of starch from starch-rich products.[16,19,26] Further, when cereal products and potatoes are extracted with hot neutral detergent solution, the starch forms a gelatinous material which impedes filtration and washing. The filtration aids suggested by Van Soest and Wine[20] do not always solve these problems. Some improvement is obtained by doubling the amount of detergent solution used. The filtered NDF in these cases gives a strong positive iodine test for starch, and only a proportion of the residual starch can be removed with 90 % aq. DMSO (Selvendran and DuPont, unpublished results). Starch can be removed from the concentrated samples (or residues) by enzymatic action, thereby aiding filtration;[27-30] such residues are referred to as modified NDF. However, it should be noted that most commercial starch-degrading enzyme preparations have hemicellulase activity and thus there is some danger of losing some of the hemicelluloses. Even after treatment with starch-degrading enzymes, the filtration rate was found to be very slow with some foods. Because very few workers have reported the carbohydrate composition of NDF and modified NDF preparations from starch-rich products, we shall give some examples in Section 3, and discuss the relative merits of this procedure in the light of such studies.

Cold neutral detergents are useful solvents, but they too solubilise a small amount of cold water-soluble pectic substances (in the case of vegetables and fruits) and some β-D-glucans and arabinoxylans (from cereals).[15,23] The cold detergent-extracted residues can be easily isolated by centrifugation or filtration, and the associated starch can be quantitatively removed by extraction with 90% aq. DMSO, provided the tissue structure has been completely disrupted.

Freeze-drying the samples, after slicing, etc., is a convenient method of storing a large number of samples, and has the advantage that, unlike alcohol treatment, the constituent polymers are not subjected to co-precipitation effects. This appears to be an advantage for the quantitative removal of starch enzymatically.[24,26] However, the procedure involving the use of freeze-dried samples suffers specifically from: (1) oxidation of the phenolics to give brown products, which would bind to polymers, particularly proteins, and thus interfere with lignin determination; and (2) possible degradation of DF polymers of fresh products by enzymatic action, because the cell wall degrading enzymes are not inactivated.[31,32]

2.2. Determination of the Monomeric Composition

2.2.1. General Comments on the Hydrolysis of Polysaccharides

The wide range of DF polysaccharides, which are composed of a variety of monosaccharides linked by glycosidic bonds of varying acid lability, precludes a single satisfactory procedure that will cleave every linkage and give each component in a quantitative yield. Further, all methods of hydrolysis cause some degradation of the constituent monosaccharides, and the stability of the sugars to hot acids varies.[33] However, depending on the carbohydrate composition of a polysaccharide, reasonably satisfactory hydrolysis conditions may be devised based on the following considerations: (1) furanoside (and deoxypyranoside) linkages are more readily hydrolysed than pyranoside linkages; (2) pentopyranosides hydrolyse somewhat faster than hexopyranosides; (3) α-glycosidic bonds are usually more labile than β; (4) with pyranoside residues, resistance of glycosidic links to acid hydrolysis increases in the order $1 \rightarrow 2 < 1 \rightarrow 3 < 1 \rightarrow 4$; (5) uronic acids linked to neutral sugars yield uranosyl-sugar disaccharides (aldobiouronic acids), which are resistant to hydrolysis; (6) some insoluble polysaccharides such as cellulose, long-chain mannan or xylan require pretreatment with cold 72% H_2SO_4 before they can be (completely) hydrolysed by 1 M acid; (7) pectins (and pectic acids) are incompletely hydrolysed by dilute acids, and this is partly due to the fact that they tend to precipitate in acidic medium, and can be dissolved only after protracted heating. Solubilising the pectic substances by treatment with polygalacturonase prior to acid hydrolysis facilitates the quantitative release of the neutral sugars. For a discussion of the above factors see refs. 33–37. Some of these aspects can best be illustrated by considering the relative stabilities of the various linkages in the A3(S1) polysaccharide from *Acetobacter aerogenes*[38] and the cell wall material of (deseeded) mature runner bean pods and potatoes.

The A3(S1) polysaccharide is made up of repeating sequences of the tetrasaccharide shown below:

$$\begin{array}{l} \qquad\quad \text{Glc} \\ \qquad\quad\ 1 \\ \qquad\quad\ \downarrow \beta \\ \qquad\quad\ 4 \\ \longrightarrow 6\text{-D-Glc } 1 \xrightarrow{\beta} 4\text{-D-GA } 1 \longrightarrow 3\text{-L-Fuc } 1 \longrightarrow \end{array}$$

The approximate ratios of the first-order constants for hydrolysis of the various glycosidic linkages in the polysaccharide are as follows:

$$\text{Fuc } 1 \longrightarrow 6 \text{ Glc:Glc } 1 \xrightarrow{\beta} 4 \text{ Glc:Glc } 1 \xrightarrow{\beta} 4 \text{ GA:GA } 1 \xrightarrow{\alpha} 3 \text{ Fuc}$$
$$= 1300:130:50:4$$

The GA → Fuc bond is extremely stable to acid hydrolysis. In fact the L-fucopyranosyl bond is hydrolysed 325 times faster than the D-glucosyluronic acid (GA) bond in 0·5 M sulphuric acid at 100 °C. This situation arises to a greater or lesser extent for all heteroglycans. The order of stability to acid hydrolysis (0·5 M H_2SO_4 at 100 °C) of the different glycosidic linkages in the polysaccharide is similar to that observed for hexosides, hexuronosides and 6-deoxyhexosides.[39] The $t_{1/2}$ for acid hydrolysis of the fucosyl bond is less than 10 min and for the glucuronosyl bond 47 h. The stability of the uronic acid bond permits recovery of almost 50% of the polysaccharide as the aldobiouronic acid after a 4–5 h

TABLE 3

COMPARISON OF HYDROLYSIS CONDITIONS

(Results expressed as μg anhydro-sugar/mg dry material)[a]

Sugar	*CWM of potatoes*				*CWM of mature runner beans*		
	2 N TFA	*0·5 M H_2SO_4*	*1 M H_2SO_4*	*Saeman*[b]	*2 N TFA*	*1 M H_2SO_4*	*Saeman*[b]
Rha	13·4	9·4	18·8	19·5	12·8	12·7	11·3
Ara	44·9	44·5	46·2	47·1	25·2	28·2	25·2
Xyl	13·2	13·4	16·5	16·2	51·7	61·8	64·8
Man	T	T	T	6·2	19·7	14·9	31·0
Gal	275	266·0	278·1	262·5	65·7	66·8	64·0
Glc	12·4	8·4	11·3	314·5	36·2	33·9	291·7

[a] Sugars released on hydrolysis for either 2 h at 120 °C (TFA) or 2·5 h at 100 °C (H_2SO_4 and Saeman hydrolysis). In Saeman hydrolysis the material is first dispersed in 72 % (w/w) H_2SO_4 to solubilise the polysaccharides, including cellulose, and then diluted to 1 M and hydrolysed. T, Trace.

[b] Molarity of diluted acid is 1.

hydrolysis period. This marked stability of the glucuronosyl bond precludes the complete and quantitative hydrolysis of the aldobiouronic acid to its monosaccharide components without extensive degradation of the sugars. This point has to be borne in mind when hydrolysing rhamnogalacturonans and glucuronoxylans with acids.

Polysaccharides are usually hydrolysed with sulphuric acid, either by direct refluxing with dilute acid, or after preliminary dissolution in concentrated acid, prior to dilution and hydrolysis.[33,37] Nitric acid is not commonly used for hydrolysis but, in combination with urea, has been recommended for pectic polysaccharides[40] and has been used in a study of apple pectin.[41] Hydrochloric acid is seldom used for hydrolysis of polysaccharides because it causes more degradation than sulphuric acid.[35] Hydrochloric acid is, however, commonly used for hydrolysis of glycoproteins; hydrolysis of such compounds may be facilitated by prior solubilisation with proteolytic enzymes.[42] Trifluoroacetic acid (TFA, 2 M) has been used for hydrolysing cell wall polysaccharides,[43] and the yields of neutral sugars from non-cellulosic polysaccharides are comparable with those obtained with dilute sulphuric acid.[25,33] TFA has the advantage that it is volatile, and can therefore be readily removed by evaporation. Hough *et al.*[44] have compared the use of H_2SO_4, HCl and TFA for hydrolysis of polysaccharides, and have reported that HCl gives lower recoveries than the other two acids.

We have compared the relative effectiveness of TFA, H_2SO_4 and Saeman hydrolysis conditions for the release of neutral sugars from the cell wall material of potatoes and mature runner bean pods; the results are summarised in Table 3. From these results and those reported by Selvendran *et al.*,[33] Selvendran and DuPont[16] and Albersheim *et al.*,[43] the following conclusions may be drawn: (1) Comparable amounts of neutral sugars are released from the non-cellulosic polysaccharides of potatoes by 2 M TFA hydrolysis for 2 h at 120 °C, and by 0·5 M and 1 M H_2SO_4 hydrolysis for 2·5 h at 100 °C. However, the amount of xylose released by TFA hydrolysis from the hemicelluloses of mature runner bean pods was appreciably lower. (2) Using 1 M H_2SO_4 hydrolysis, the yield of neutral sugars from the cell wall preparations was not appreciably increased by periods of hydrolysis $>2{\cdot}5$ h, except for rhamnose and glucose. The slow release of rhamnose and glucose is probably due to the stability of the aldobiouronic acid (GalA 1 $\rightarrow$ 2 Rha) and the insolubility of cellulose in the dilute acid. The yield of rhamnose after 2·5 h of hydrolysis was only 60 % of that after 5 h. (3) Saeman hydrolysis conditions hydrolysed the cellulose almost completely in 2·5 h, and there was a noticeable increase in the level of mannose and xylose as well. Longer periods of hydrolysis degraded the liberated sugars.

Hough *et al.*[44] have reported that the degradation of the sugars can be minimised by conducting the hydrolysis in a nitrogen atmosphere. The degradation of the sugars can also be minimised by using cation-exchange resins in the protonated form, in very dilute acids, for hydrolysis.[45] This method is, however, suitable only for soluble polysaccharides and is becoming increasingly popular in the field of glycoproteins. In this connection, it is useful to note that water-soluble polystyrene sulphonic acid has been used in the controlled fragmentation of labile polysaccharides.[46] Some workers have performed hydrolysis in an autoclave, but it appears that 0·5 M H_2SO_4 at 120 °C causes appreciable degradation of L-arabinose and D-galactose.[47]

An alternative method for cleavage of polysaccharides is by methanolysis;[48] neutralisation of the acid after methanolysis is done by the addition of silver carbonate. Methanolysis protects the liberated reducing group as the methyl glycoside and thereby reduces the possibility of unwanted side reactions. At the same time the carboxyl groups of uronic acids are converted to the methyl esters which are more stable. Monosaccharides are generally stable for 24 h in methanolic 1 M and 2 M HCl at both 85 and 100 °C, but undergo considerable destruction in methanolic 4 M and 6 M HCl at 100 °C. Since methanolysis, in 1 M methanolic HCl, appears to be as efficient as acid hydrolysis at cleaving glycosidic linkages, and causes less destruction of sugars than does aqueous acid, the possibility of using it for soluble polysaccharides should be considered.

The sulphuric acid hydrolysates are neutralised, usually with $BaCO_3$ or $Ba(OH)_2$, although ion-exchange resins and methyldioctylamine have also been used.[44,49] Hough *et al.*[44] have found that neutralisation with methyldioctylamine showed the least loss. However, the error due to selective losses, by adsorption on the precipitate, can be minimised by including an internal standard after the neutralisation step.

In order to obtain accurate results, correction factors must be established for the decomposition of each monosaccharide under the conditions of hydrolysis used. Also, allowance must be made for incomplete hydrolysis of polysaccharides containing uronic acid residues. In fact, it has been shown by several workers that the common aldobiouronic acid 2-*O*-(4-*O*-methyl-α-D-glucopyranosyluronic acid)-D-xylose, from 4-*O*-methylglucuronoxylans and acidic arabinoxylans, accumulates in the hydrolysate and is relatively resistant to further hydrolysis.[50–52] This accumulation of aldobiouronic acid is responsible for some apparent loss of xylose in the final analysis.

In this connection, it is useful to note that several factors influence the quantitative determination of uronic acids. These include: incomplete hydrolysis, decarboxylation caused by the acid hydrolysis conditions used, the problem of lactonisation (and thus incomplete removal from the neutral sugars), losses by adsorption on the $BaSO_4$, and the difficulties encountered in the formation of suitable volatile derivatives for analysis by GLC. Some of these problems have been studied by Blake and Richards and some measures for overcoming them have been suggested.[53,54] A useful method for water-soluble polyuronides is to convert the uronosyl residues to the corresponding hexosyl residues, by reducing the carbodiimide-activated carboxyl groups with $NaBH_4$. This procedure replaces the acid-resistant uronosyl-sugar bonds with the more acid-labile glycosyl bonds.[55] The resulting neutral polysaccharides can then be analysed by formation of the corresponding alditol acetates. By performing the reduction with $NaBD_4$ the hexosyl residues, arising from the uronosyl residues, can be dideutero-labelled which can be distinguished from the unlabelled hexosyl residues by combined GC–MS. This method has been used successfully for soluble pectins,[56] but may not be suitable for DF preparations which contain both water-soluble and water-insoluble polyuronides.

Since all the methods of hydrolysis have some limitations and cause variable destruction of the different sugars, a compromise must be made between maximum hydrolysis and minimum destruction.

2.2.2. Analysis of Monosaccharides

(*a*) *Reducing Sugar and Colorimetric Methods*

Once the sugars have been released by the hydrolytic methods discussed above, it should, in principle, be possible to determine the total number of sugar residues by estimation of the reducing power. There are several techniques available for this purpose.[57–59] The two most commonly used methods involve the reduction of alkaline copper salts and alkaline ferricyanide. These methods suffer specifically from the fact that different sugars have different reducing powers in alkaline solution. The major factors that affect the reaction are the rate of heating, the alkalinity and the strength of the reagent; these factors may cause varying alkaline degradation of the sugar. A further difficulty arises from the fact that the side-chains of some amino acids, e.g. tryptophan, tyrosine and serine, exhibit reducing properties.[60] These amino acids (or peptides containing them) could arise from intracellular proteins present in the DF preparations. Therefore the accuracy of the method for determining sugars in mixtures is largely dependent on the degree to which the sugars can be separated from each other and from interfering components.

Alternative methods for sugar estimation are based on the formation of coloured derivatives; these methods do not require a prior release of the sugar as a separate step. Colorimetric determination of sugars involves two steps: (a) the formation of a chromogen from the sugar, and (b) the condensation of the chromogen with a specific chromophore to produce a coloured derivative. Most methods are based on the reaction of the sugar with an acid to give a derivative of furfural, which gives a coloured product with an appropriate phenol or aromatic nitrogen base. Some of the commonly used methods are described by Dische.[61] Problems are likely to arise because different sugars give different colour yields. Because of the relatively low specificity of some of these colour reactions, it is often important to apply them only after appropriate separation of the constituents.

For example, with the anthrone method for hexoses, the colour yields of galactose, mannose, glucose and fucose vary. The approximate molar colour yields under standard conditions of the reaction, compared to galactose, are: galactose, 100; mannose, 96; glucose, 162; and fucose, 60.[62,63] We have found that galactose gives about half the response of glucose,[64] and that a suitable correction factor for the hexoses cannot be introduced without prior knowledge of the sugars in the mixture.

Similarly, there are problems associated with the estimation of pentoses. Although hexoses do not interfere with the estimation of pentoses by the phloroglucinol method when the ratio of hexose to pentose is $<10:1$, the interference becomes quite significant at higher ratios of hexose to pentose.[64] Glucuronic acid and galacturonic acid, on the other hand, give about 50% and 25% respectively of the response of xylose at equimolecular concentration (Selvendran and DuPont, unpublished results).

Several factors have to be borne in mind when estimating the uronic acid content of DF preparations by the colorimetric methods. These are: (1) incomplete hydrolysis of polyuronides; (2) sensitivity of the uronic acid to strong acids; (3) the interference from hexoses, pentoses and various phenolic compounds—the real problem arises when the concentration of uronic acid is very low compared with the amount of hexoses and pentoses (e.g. fibre from cereals and cell wall material from lignified tissues); (4) the different responses from the different types of uronic acids: galacturonic acid from pectins, glucuronic acid from hemicelluloses and others from food additives. In alginates, for example, the colour yields of guluronic and mannuronic acids are quite different.[65]

In view of these difficulties, it is suggested that uronic acids are best estimated by the decarboxylation method. This method depends on the quantitative release of CO_2 from uronic acids and their derivatives on heating with hydroiodic acid, and has been used for DF preparations by Theander and Aman.[18]

(*b*) *Separation, Identification and Determination of Individual Sugars*

(*i*) *Paper chromatography.* Although paper chromatography is not used widely nowadays for the separation and estimation of sugars, some pertinent comments will be made because the technique is sufficiently reliable if specialised equipment is not available. The sugars resulting from mineral acid hydrolysis may be separated on paper using a number of solvents.[66] It is necessary to remove the mineral acid prior to chromatography. The most commonly used solvents for neutral sugars are mixtures of butanol (or ethyl acetate), pyridine and water, and for acidic sugars, ethyl acetate, acetic acid and water.[66,67] The use of a marker is often necessary because the solvent front is usually allowed to run off the paper to ensure maximum separation. The sugars are revealed as coloured spots after a suitable spray treatment. It is useful to note that unless special precautions are taken, uronic acids can exist in un-ionised and ionised forms, giving either double spots or long tails, particularly in basic solvents. The separated sugars can be quantitatively estimated, after elution from the paper, by reducing procedures such as alkaline ferricyanide (1–9 μg) or by colorimetric methods.[68–70] The application of this method for the analysis of sugars from cell wall material from a range of plant tissues has been described.[40,71,72]

Since glucose and galactose are not well separated by paper chromatography in most solvents, it is advisable to run the chromatograms for an extended period of time. If the amounts of glucose and galactose present in a sample do not permit adequate separation of the sugars, the entire spot may be eluted. One aliquot of this eluate may be analysed for glucose enzymatically, and another aliquot may be analysed for the sum of glucose and galactose by a reducing procedure. The value for galactose may then be obtained from the difference.

(*ii*) *Anion-exchange chromatography.* The borate complexes of neutral sugars can be separated by anion-exchange chromatography and a number of procedures have been reported.[44,73–77] In our laboratory, the automated method of Davies *et al.* was used for the separation of neutral (and acidic) sugars from the cell wall material of runner beans[78,79] and potatoes,[33] and the method has given values in close agreement with those obtained by GLC. The method has the advantage that the mixture of sugars can be analysed directly without derivatisation, and automation of analysis can be easily accomplished. Resolution of the sugars under appropriate conditions can be as good as or even better than GLC. It should be noted that since the molar colour yield of each sugar varies with the detection system used, suitable colour adjustment factors must be introduced when calculating the amount of sugar under each peak.

(*iii*) *Gas–liquid chromatography*. Most sugars are not sufficiently volatile to be used for GLC, and they must therefore be converted into suitable volatile derivatives. Reviews have been published which describe the utilisation of such derivatives for the separation of sugars.[37,80] The most commonly used derivatives of the reducing neutral sugars have been methyl ethers, trimethylsilylethers of the free sugars or their methyl glycosides, acetates and trifluoroacetates. Though the above derivatives fulfil most of the requirements, the main difficulty in working with monosaccharides is the formation of usually two, but maybe as many as four, glycosides per monosaccharide resulting from anomeric and ring isomerisation, each of which produces a peak on the chromatogram.[81,82] This difficulty has been overcome by reducing the sugars to their alditols with $NaBH_4$,[83,84] and then separating the alditol derivatives;[85-87] each sugar after reduction gives rise to only one peak.

The same objective may be achieved by conversion into the corresponding nitrile. Reaction of a sugar with hydroxylamine in pyridine gives the oxime, which may be directly dehydrated to the nitrile and then acetylated to the acetate. The separation of aldononitrile acetates of neutral sugars by GLC has been reported.[88,89] However, since alditol acetates have been found to give the most satisfactory separation of sugar mixtures likely to be encountered in DF analysis,[16,18,90-92] this derivative is widely used. The details of the method as used in our laboratory are given in ref. 33; the recommended method offers a consolidation of the recent modifications of the alditol acetate procedure.

It should be noted that one of the great advantages of GLC is that the detection of the components depends on measurement of a physical property, and not on the reaction of a functional group. Therefore GLC methods are not subject to the errors encountered in colorimetric techniques.

(*iv*) *High-performance liquid chromatography* (*HPLC*). Recent developments in HPLC equipment and various types of micro-particulate column packings offer considerable promise for the direct and rapid determination of sugars, including oligosaccharides. In HPLC, as in GLC, the separation of the components of a mixture is based on the characteristic retention times within the column of the instrument. As the packing material is very fine, high pressures have to be employed. Developments in improved pumping systems capable of maintaining a constant liquid flow rate over a period of time and improved columns capable of separating closely related sugars in a mixture have made HPLC a valuable tool over the past few years. The resolved components are usually detected by a differential refractometer whose output is fed to a strip-chart recorder.

Three systems have been commonly employed for the separation of sugars: (1) Bondapak AX/Corasil with acetone/ethanol/water (20:2:1) as the eluant, has been used to resolve rhamnose, xylose, fructose, mannose and galactose. The separation of galactose and glucose has proved difficult on this column.[93] Linden and Lawhead[94] have tried both Bondapak AX/Corasil (35–50 μm particle size) with ethyl acetate/isopropanol/water (25:50:35) as eluant, and (2) Bondapak Carbohydrate phase (10 μm particle size) with acetonitrile/water (75:25) as eluant, for the separation of sugar mixtures (fructose, glucose, sucrose and raffinose), and have obtained better resolution with the latter system. The Bondapak Carbohydrate phase is amino-bonded (alkylamine-modified silica) and has the advantage of being resistant to high pressures enabling the use of high eluant flow rates. The separations can be carried out at room temperature. Both packing materials are products of Waters Associates. The Bondapak Carbohydrate column with acetonitrile/water (85:15) as eluant is, however, not suitable for separating sugars from cell wall hydrolysates.[95] In this system the peaks corresponding to rhamnose, xylose and arabinose can be readily resolved, but those of mannose, glucose and galactose tend to overlap. In our laboratory also, similar difficulties were experienced with Bondapak Carbohydrate column. (3) The third system is a cation-exchanger (Ca^{2+}) with pure water as the eluant.[96,97] In the Ca column system, the main function of the ion-exchanger is to immobilise Ca^{2+}, and the separation results from the different complexing abilities of the sugars with Ca^{2+}. The degree of cross-linking of the resin is an important factor which determines the separation of higher oligosaccharides. For the separation of higher oligosaccharides, resins with a low degree of cross-linking must be used.[96] The cation-exchange resin (Aminex HPX-87P), marketed by Bio-Rad Laboratories, uses water as the eluant at 85 °C, and appears to separate most of the sugars from cell wall hydrolysates (glucose, xylose, galactose, arabinose and mannose). This system, or an improved version of it, may prove to be very useful for the separation of sugars from DF preparations.

For the separation of uronic acids (D-galacturonic acid, D-glucuronic acid, L-guluronic acid and D-mannuronic acid) from DF preparations, the HPLC system proposed by Voragen *et al.*[98] could be used. In this procedure, the uronic acids are separated on a strong anion-exchange column (Nucleosil 10SB or Zorbax SAX) using sodium acetate buffers as eluant. This method does not separate D-galacturonic acid and L-guluronic acid, but this should not cause any real problems, as these uronic acids are present in very different types of DF preparations.

2.3. Determination of Lignin

The quantitative determination of lignin, in the tissues of plants used as foods, is beset by a number of difficulties. These arise from: (1) uncertainty of the detailed chemistry of lignin; (2) differences in the chemical nature of materials associated with lignin; for example, lignin is apparently linked to some hemicelluloses by covalent bonds and so is difficult to separate from these polysaccharides; (3) the differences in the solubility characteristics of lignin in acid; (4) the chemical reactions of lignin with accompanying materials; and (5) artefacts produced during food preparation. Hence, there is at the present time no entirely satisfactory method for the quantitative determination of lignin. For a critical review of most of the methods used for lignin determination see Browning.[99]

Although a great deal of work has been carried out to elucidate the factors affecting the quantitative determination of lignin, much more remains to be done. This is because many modifications in procedure have been made, possibly without full realisation of the interacting factors and sources of error. The principles underlying lignin determination and the important limitations, which must be borne in mind in the interpretation of the results, will be outlined using suitable examples. A comparison of 'apparent lignin' values, obtained by various methods with comparable samples, is necessary, because in many instances the methods have been adapted and applied indiscriminately and uncritically.

The most generally used methods for lignin determination are: (1) the sulphuric acid hydrolysis method;[99,100] (2) the permanganate oxidation methods;[22,99,101] and (3) the acetyl bromide method.[102–104]

2.3.1. The Sulphuric Acid Hydrolysis Method

This method is based on the hydrolysis and solution of the accompanying polysaccharides with H_2SO_4, leaving a residue which after washing and drying is weighed as lignin. The use of H_2SO_4 for determining lignin was first applied by Klason and, although the original method has been modified extensively, the lignin isolated by this method is usually referred to as 'Klason lignin'. The hydrolysis of the polysaccharides is accomplished effectively by pretreatment with cold 72% (w/w) H_2SO_4 (to solubilise cellulose, long-chain xylan or mannan) and then diluted with water and boiled to complete the hydrolysis. Suitable pre-extractions are necessary to remove extraneous material that would interfere with the assay. Examples are sugars, lipids, polyphenols and (degradation products of) proteins which may condense with the lignin during the acid treatment and erroneously appear as lignin. The protein interference is particularly serious in the analysis of immature tissues (certain vegetables) and cereal products, in which the protein content is high and the degree of lignification is low. For such products, removal of proteins by treatment with proteolytic enzymes, or detergents, or hydrolysis with dilute acids, is recommended. However, pre-extractions drastic enough to remove the major portion of interfering materials undoubtedly lead to some loss of lignin. The most commonly used procedure for Klason lignin determination is the Tappi Standard Method.[99,100] Corrections for ash are usually unnecessary, except for grasses which contain significant amounts of silica.[105]

It is useful to note that for conifer woods, which contain predominantly guaiacyl lignins, there is good correlation between Klason lignin and total lignin. In some species of wood, unextractable polyphenols may cause a positive error by condensing with lignin. With angiosperm woods, which contain a fair proportion of syringyl lignins, treatment with 72% H_2SO_4 results in 25–60% of the total lignin going into solution, and the dissolved lignin consists predominantly of syringylpropane residues. Dilution with water (and hydrolysis) precipitates most, but not all, of the dissolved lignin and part of it passes into the filtrate as 'acid-soluble lignin'. With beech wood about 50% of the lignin is acid-soluble. Herbaceous plants including monocotyledonous species, like angiosperm woods, also contain a fair proportion of acid-soluble lignin.[14] Some errors may also stem from partial hydrolysis of *p*-coumaric and ferulic ester groups.

2.3.2. The Permanganate Oxidation Methods

Being phenolic in nature, the aromatic nuclei of lignin are quite susceptible to the oxidative attack of a variety of reagents. Potassium permanganate in acidic solution, for example, oxidises lignin readily, degrading it to CO_2 and dibasic acids; the associated carbohydrates are also (thought to be) partially degraded.[106] Two methods of lignin determination which involve treatment with $KMnO_4$ are as follows. In one, the consumption of permanganate under standardised conditions is taken to give an estimate of lignin. The method is based on the addition of an excess of $KMnO_4$ to the sample and determining by back-titration the amount consumed. The reaction is terminated by the addition of potassium iodide (or an excess of ferrous sulphate solution). The permanganate number (or kappa number) is usually expressed as the amount of 0·1 N $KMnO_4$ solution consumed by 1 g of pulp. The permanganate lignin content (permanganate number × 0·15) of wood samples has been found to be approximately equal to the Klason lignin values. However, it should be noted that this procedure is not widely applicable, because the permanganate number is influenced by several factors, such as amount and concentration of $KMnO_4$, temperature and reaction time, and by the type of phenolic residues present in the lignin.[99,100]

In the other method, the (partially degraded) lignin is removed from the acid detergent residue by treatment with $KMnO_4$ and lignin is obtained from the difference in weights.[22,101] Application of this method to whole cell wall material of vegetable and cereal products is unsatisfactory. This is because some of the hemicelluloses associated with the lignin by covalent linkages, or held in the cell wall matrix by phenolic ester cross-linkages, are also solubilised and thus removed, giving elevated values for lignin (Selvendran and DuPont, unpublished results).

2.3.3. The Acetyl Bromide Method

This method is based on the spectrophotometric determination of lignin as the acetylated derivative, and was first applied to wood samples.[99,100,102] The sample is treated with a mixture of warm acetyl bromide and acetic acid, which results in the acetylation of the lignin and associated polysaccharides, thus rendering them soluble in the medium. Some of the polysaccharides may be degraded by the treatment. After destruction and removal of excess reagent and precipitated proteins, the absorbance of the resulting solution is read at 280 nm.The contribution of the acetylated polysaccharides to the absorption is usually very small and can therefore be ignored. Absorption at 280 nm was found to be proportional to the lignin content of the samples, as determined by other methods, but the particle size of the samples was critical for obtaining satisfactory spectra. With minor modifications, this method has been used by Morrison for grasses, hays, silages and legumes.[103,104]

Although there is a linear relationship between the acetyl bromide lignin and Klason lignin, for different weights of lignified tissues from the same plant (e.g. deseeded pods, parchment layers and strings from mature runner beans), the gradient and intercept vary with different materials (Selvendran and DuPont, unpublished results). This makes it difficult to use one standard graph to calculate the lignin content from a range of products. In Morrison's work,[103] the comparison is between different organs from the same plant or comparable species.

2.3.4. Comparison of Methods for the Determination of Lignin

Because the nature of the phenolic residues, structure and molecular weight of lignins from cereal products and vegetables are not fully known, it is difficult to decide which of the analytical methods for lignin determination gives the most meaningful results. In order to assess the relative merits of the various methods, the lignin values from a range of products were compared; the results are given in Table 4. As can be seen from the table, there is considerable variation in the lignin values obtained by the different methods. However, some general conclusions can be drawn if the materials are considered separately.

TABLE 4
COMPARISON OF METHODS USED FOR LIGNIN DETERMINATION
(Results are expressed as % dry wt of the material analysed)

Material analysed	*Klason lignin*	*$KMnO_4$ lignin*			*Acetyl bromide lignin*
		Van Soest gravimetric	*Kappa lignin*	*Titrimetric*[a]	
Fine-milled bran	5·8	77·2	8·3	4·6	7·1
CWM bran	9·3	75·6	10·6	8·9	7·2
ADR bran[b]	26·3	25·5	7·3	4·1	2·2
Fine-milled oat hulls	19·2	29·7	n.d.	n.d.	10·3
CWM parchment layers of runner bean pods	16·7	54·5	20·1	11·2	13·5
ADR mature runner beans (ex AIR)[c]	10·0	33·9	6·7	3·7	6·8

[a] Values calculated based on tannic acid as standard.
[b] Yield ADR from original tissue 32·4%.
[c] Yield ADR from AIR 41%.
n.d., not determined.

(a) Cell Wall Material of Wheat Bran

The CWM of wheat bran was prepared by sequentially extracting the fine-milled bran with aq. 1% Na deoxycholate, phenol/HOAc/H_2O (2:1:1, w/v/v) and aq. 90% DMSO. With the CWM there is good agreement between the values for Klason lignin, permanganate lignin (titrimetric) and acetyl bromide lignin, although the value for permanganate lignin (gravimetric) is very high. A comparison of the carbohydrate compositions of the CWM and $KMnO_4$-insoluble residue showed that an appreciable proportion of the arabinoxylans were solubilised along with the lignin during the permanganate treatment. The arabinoxylans are held in the cell wall matrix by phenolic ester cross-links (although some of them may be covalently linked to the lignin) so, on oxidation of the phenolic esters and lignin, the matrix polysaccharides are also solubilised giving elevated values for lignin. From these results, and those of the alcohol/benzene-extracted fine-milled oat hulls, it would appear that the $KMnO_4$ oxidation method (gravimetric) is probably unsuitable for whole cell wall preparations.

(b) Acid Detergent Residue (ADR) of Wheat Bran

The ADR of bran was prepared from the fine-milled bran by extraction with hot cetyltrimethylammonium bromide in diluted H_2SO_4 as described by Van Soest.[107] With this ADR there is good agreement between values for Klason lignin and $KMnO_4$ lignin (gravimetric), but these values are considerably higher than those for $KMnO_4$ lignin (titrimetric) and acetyl bromide lignin. During extraction with hot acid detergent, a fair proportion of the acid-soluble lignin and phenolic ester residues are hydrolysed, in addition to the arabinoxylans, and thus removed from the ADR. Also some of the phenolic residues of the lignin are modified and these are probably not readily acetylated, and would account for the very low values for acetyl bromide lignin.

(c) CWM of Parchment Layers of Runner Bean Pods

CWM was prepared from parchment layers carefully dissected from mature runner bean pods. This preparation contains mainly cellulose, xylans and lignin in the approximate ratio 1:0·7:0·8; for details of analysis see ref. 108. With this product there was reasonable agreement between the values of Klason lignin and acetyl bromide lignin. As expected, the value for $KMnO_4$ lignin (gravimetric) was high because some of the xylan (and possibly cellulose) was solubilised along with the lignin.

(*d*) *Acid Detergent Residue of Mature Runner Beans*
With this product the Klason lignin value was considerably lower than that for the $KMnO_4$ lignin (gravimetric), but compared well with the acetyl bromide lignin value. Hence, very little is gained in preparing the ADR from a product and then determining its $KMnO_4$ lignin (gravimetric) content. Despite their limitations, the Klason lignin and acetyl bromide lignin values of the whole cell wall material give a reasonable measure of lignin content. The acetyl bromide method is to be preferred for the dietary fibre, because it can be scaled down for very small samples (5–10 mg).

3. ISOLATION AND ANALYSIS OF MILLIGRAM QUANTITIES OF DIETARY FIBRE: AN ASSESSMENT OF THE DIFFERENT METHODS

3.1. Detergent Extraction Procedures of Van Soest *et al.*[20–22]
The neutral detergent fibre (NDF) method is well documented; for practical details see Van Soest and Wine[20] and Robertson and Van Soest.[21,22] The method depends on the property of the hot aqueous detergent to solubilise intracellular proteins (and some starch). It essentially consists of refluxing the fresh or air-dried material with neutral detergent solution, containing 0·5% (w/v) sodium sulphite, for 1 h. The mixture is then filtered, washed with hot water and acetone, dried and weighed. Ashing the residue and reweighing gives an estimate of the ash-free NDF. One litre of the neutral detergent solution contains 30 g sodium lauryl sulphate, 18·61 g disodium salt of EDTA ($Na_2EDTA.2H_2O$), 6·81 g sodium borate ($Na_2B_4O_7.10H_2O$), 4·59 g disodium hydrogen phosphate (anhydrous) and 10 ml ethylene glycol, pH range 6·9–7·1. The main problems encountered in this type of analysis are: (i) loss of lignin (subunits) by the action of hot sodium sulphite solution;[109] (ii) high levels of starch impede filtration rates considerably, and the NDF is heavily contaminated with starch (modified?), giving an overestimate of neutral detergent fibre; (iii) loss of detergent-soluble arabinoxylans, arabinogalactans and β-glucans from cereal products; and (iv) loss of an indefinite amount of pectic substances from vegetables and fruits.

3.1.1. Suggested Modifications for Starch-Rich Foods
(1) In the first modification, the residual starch is removed by the action of a heat- and detergent-stable α-amylase from *Bacillus subtilis* (e.g. α-amylase, type IIIA, Sigma A6505; optimum activity at pH 6·9 and

TABLE 5

ANALYSIS OF NDF, MODIFIED NDF AND ENZYME-TREATED AIR OF SOME FOOD PRODUCTS

	Oats			Brown bread		Potatoes			Apples	
	NDF	*M-NDF*	*E-AIR*	*M-NDF*	*E-AIR*	*NDF*	*M-NDF*	*E-AIR*	*NDF*	*AIR*
Yield[a] (gravimetric)	6·5	5·6		5·2		1·5	1·1		1·1	
Yield[b] (carbohydrate)	3·1	2·1	6·7	3·3	5·5	1·5	0·95	2·1	1·1	1·8
Anhydro-sugar[c] (μg/mg product):										
Deoxy-hexose	1·7	1·6	5·3	—	1·1	3·1	5·4	7·8	28·9	18·8
Ara	71·4	76·0	55·7	155·1	52·1	15·2	10·8	30·4	79·3	109·9
Xyl	105·2	116·2	80·3	242·7	78·3	4·0	8·6	12·9	83·8	48·7
Man	5·0	8·9	9·7	7·0	7·7	6·3	4·6	2·9	28·4	18·5
Gal	7·3	9·6	11·8	9·3	7·7	27·3	38·4	128·7	63·3	53·7
Glc	264·3	128·0	108·7	195·0	91·3	934·2	783·1	147·3	505·8	268·2
	(155·3)	(54·3)	(52·3)	(58·2)	(42·3)	(489·7)	(33·1)	(55·1)	(36·1)	(14·1)
Uronic acid	26·9	31·1	24·4	19·6	20·6	46·2	48·0	128·6	176·7	276·2

[a] g/100 g product as bought.
[b] Yield of DF based on carbohydrate analysis of NDF, M-NDF and E-AIR preparations.
[c] Sugars released on Saeman hydrolysis (figures in parentheses give glucose released on 1 M H_2SO_4 hydrolysis).

40 °C).[21,29,110] As the enzyme is heat-stable, it is added to the food product dispersed in hot detergent solution. The detergent (apparently) inactivates the enzyme which hydrolyses the $\alpha(1 \rightarrow 6)$ glucosidic linkages but not the one which hydrolyses the $\alpha(1 \rightarrow 4)$ linkages.[21] The mixture is filtered, treated with additional enzyme solution on the filter, washed with boiling water, acetone, and dried.

(2) In the second modification,[28,111] the residual starch is removed by treating the neutral detergent residue with α-amylase (Sigma A6880) on the filter funnel; the enzyme is claimed to degrade modified starch as well. The enzyme-treated residue is washed with distilled water, acetone, and dried. The modified NDF methods are usually regarded as being reasonably satisfactory for cereal products.

The first modification overcomes, to a certain extent, the problems encountered in filtering extracts from starch-rich products. The second modification, however, does not offer any effective solution for filtration problems, and may be suitable only for products which are not very rich in starch, e.g. wheat bran and bran-based products. The authors who have recommended both modifications rely on gravimetric measurements for dietary fibre values, and have not published any carbohydrate analyses of NDF and modified NDF.

As it would be difficult to comment on the relative merits of these methods without a knowledge of the carbohydrate compositions of the preparations, we have compared the composition of NDF and modified NDF, from a range of products, with those obtained from the enzyme-treated alcohol-insoluble residues of the products.[16] The products examined were brown bread, dehulled oats, potatoes and apples. The yields and carbohydrate compositions of the various preparations and their dietary fibre values, based on gravimetric measurements and carbohydrate analysis, are given in Table 5. The modified NDF values (gravimetric) for oats and brown bread are 15% and 6% less than the corresponding DF values obtained from carbohydrate content of the enzyme-treated alcohol-insoluble residues. Normally one is inclined to regard these differences as being relatively small and to arise from the solubility of some of the DF polymers in the hot neutral detergent solution. However, analysis of the modified NDF preparations shows that carbohydrates account for only 61% (brown bread) and 35% (dehulled oats) of the dry weight of the modified NDFs. These analyses query the validity of the modified NDF methods even for cereal products and indicate the need for further studies.

Since appreciable amounts of pectic polysaccharides are solubilised by the hot neutral detergent solution from fruits and vegetables, the values for the modified NDF would be considerably lower. In the case of potatoes, it appears that there is some modified starch in the modified NDF preparation.[16]

Thus it appears that the usefulness of the rapid detergent gravimetric method is offset by a number of problems arising from solubility of the DF polymers in the extraction medium, and the incomplete removal of starch from starch-rich products, which results in slow filtration rates.

3.2. Observations and Comments on the Southgate Method

In the Southgate method, the total dietary fibre is regarded as equivalent to the unavailable carbohydrates, and the method is an extension of that described by McCance *et al.* (see ref. 17). The method for the determination of unavailable carbohydrates was first described in 1969,[112] and improved modifications of this method have been published subsequently.[17,113,114] For the purpose of this discussion, we shall consider the method as given in ref. 113, which is similar to those given in refs. 17 and 114, and the reader is advised to look up ref. 114 for details. The method is shown schematically in Fig. 1. It should be noted that in ref. 114 the gelatinisation step has been added, but we shall discuss the method as described in papers to 1979 as this is the most widely used. It is claimed to provide a measure of total dietary fibre as the sum of the component non-cellulosic polysaccharides, cellulose and lignin. The hexose and pentose contents of the polysaccharides are determined by colorimetric methods, after hydrolysis. The main advantages of the method are that it is relatively simple for the amount of information it provides, and does not require sophisticated instrumentation.

Although this method is widely used, its inherent disadvantages have often been overlooked. This situation arose because there have been relatively few comparative studies. The disadvantages may be considered to arise from co-precipitation effects, incomplete tissue disruption, incomplete pectin solubilisation, inaccurate lignin values, and lack of good correlation between sugar estimates by colorimetric methods and GLC techniques for certain products.

The problems arising from co-precipitation effects are common to all methods using methanol- or ethanol-insoluble residues, particularly for starch- (and protein-) rich products, and have been considered under Section 2.1. Although gelatinisation of starch helps in its enzymatic degradation, it appears to be incomplete with some of the starch-rich products. In a recent survey of the DF content of a range of fresh and processed foods,[114] it was found that the application of the Southgate method resulted in an overestimate of the DF content of rye flour, rye

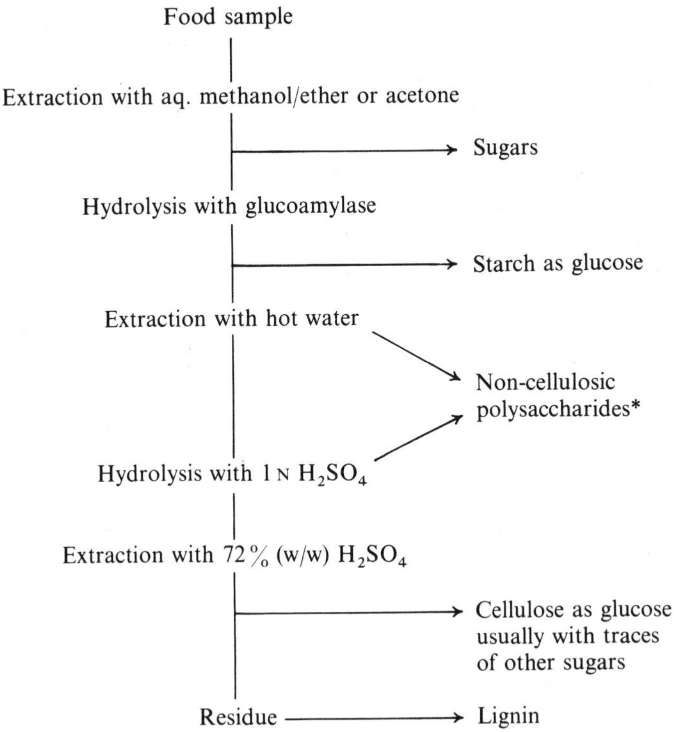

* As the component hexoses, pentoses and uronic acids.

FIG. 1. Scheme of analysis (Southgate *et al.*[17]). (Courtesy Blackwell Scientific Publications, Oxford.)

biscuit and potato powder. These discrepancies can be traced to incomplete removal of starch and inaccuracies in the estimation of sugars by colorimetric methods. In addition to co-precipitation effects, incomplete disruption of the tissues is also a factor which hinders complete removal of starch.

Extraction of the amyloglucosidase-treated residue with hot water results in the incomplete solubilisation of the pectic substances. In our experience with vegetables, further treatment of the residue with 0·5 M ammonium oxalate resulted in the solubilisation of additional amounts of pectic substances which are richer in galacturonic acid residues.[72,115,116] With cereal products, relatively small amounts of DF polymers are solubilised by hot water treatment. Therefore, the step involving extraction with hot water is of questionable value.

Since lignin is measured as the weight of the final residue, and the product has been exposed to different acid treatments, one would expect a loss of acid-soluble lignin. Further, the co-precipitated proteins and some Maillard reaction products (in the case of some processed foods), which may be incompletely solubilised by the various extraction procedures used, would also contribute to the lignin value.

The colorimetric methods for determining pentoses, hexoses and uronic acids are convenient but suffer from the fact that different sugars give different colour yields and there is usually interference by different sugars in the determinations. Some pertinent comments on these aspects have been made earlier,[23,64] and some additional data on the determination of monosaccharides and polysaccharides are discussed by Laine *et al.*[117] These workers have compared the pentose and hexose contents of DF from wheat bran and rye flour by colorimetry and HPLC. For both products the values for pentoses are in reasonable agreement, but hexoses are overestimated by the colorimetric method. We have compared the carbohydrate compositions of DF from a range of cereals and vegetables by both colorimetric and GLC methods and the implications of our findings are discussed in Section 3.3.3.

The limitations in the measurement of the different components of DF by the Southgate method should be considered when applying it to various food products. Special care should be taken to check for the presence of residual starch in the enzyme-treated residues, particularly from starch-rich products. The residual starch can be detected by the blue coloration it gives with I_2/KI.

3.3. Methods Based on the Determination of Sugars by GLC

3.3.1. Observations on the Procedure of Theander and Aman

The procedure of Theander and Aman measures the total DF as the sum of the water-soluble and water-insoluble dietary fibre fractions. The main characteristics of the method, which involve fractionating the food product after extraction with 80% ethanol and chloroform and removal of the starch enzymatically, are given in Fig. 2. The two fractions are analysed for neutral sugars and uronic acid content, the former by GLC of the alditol acetates, and the latter by a rapid decarboxylation method. For a complete description of the procedure see ref. 18.

In this procedure the water-soluble DF polysaccharides are isolated by freeze-drying the dialysed extract. This is to minimise the loss of aqueous alcohol-soluble polysaccharides, such as some arabinans and arabinogalactans.[18] The neutral sugars liberated on sulphuric acid hydrolysis of the

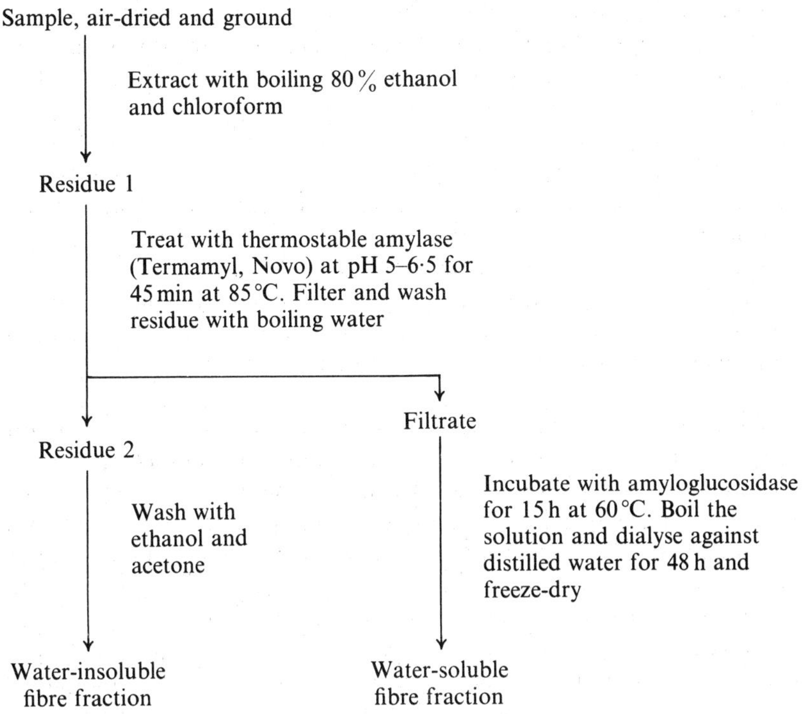

FIG. 2. Scheme of analysis (after Theander and Aman[18]).

DF fractions were determined by GLC after conversion to alditol acetates, according to Sawardeker *et al.*[85] The hydrolysis is effected by pretreating the polymers with 72% (12 M) H_2SO_4 at room temperature for 6 h.

In their first paper, Theander and Aman[18] state that starch is (almost) completely degraded in the products tested by the thermostable bacterial enzyme at 85 °C. However, in their subsequent paper[118] they recommend the incorporation of the amyloglucosidase treatment step, as a precautionary measure. The authors claim that under the experimental conditions used, Termamyl enzyme did not liberate any detectable amounts of sugars from a purified β-glucan from barley seed, a purified arabinoxylan from barley straw, or cotton cellulose. From this it is inferred that the enzyme is free of glucanase, xylanase or arabinosidase activities. However, as the work of Englyst (unpublished results) and Englyst *et al.*[24] suggests that Termamyl contains appreciable β-glucanase activity at 40 °C, it is probable that the hemicellulases are inactivated at 85 °C.

An important feature of the method is the determination of the uronic acid content in the fibre fractions by a rapid direct method. It is based upon decarboxylation with hydroiodic acid and recording the release of carbon dioxide, trapped in sodium hydroxide solution, by conductivity measurement. Contrary to the colorimetric methods, the authors claim that this method is not influenced by the neutral sugars and other constituents, which is a definite advantage. Tests with different types of uronic acid and uronic acid-containing compounds, such as D-galacturonic acid, D-glucuronolactone, aldobiouronic acid (2-*O*-(4-*O*-methyl-α-D-glucopyranosyluronic acid)-D-xylopyranose) and pectic acid, have shown they all give the same change in conductivity per mole of uronic acid. This similarity of response is very useful in the determination of the uronic acid content of DF.

The procedure of Theander and Aman has proved to be satisfactory for the determination of the DF content of a range of fresh and processed foods from cereals and vegetables. The method has been carefully checked and is to be recommended. The authors appear to claim complete removal of starch from all starch-rich products. We find it difficult to accept this conclusion for some processed foods. An aspect which we thought warranted further investigation was the possible degradation of pectic substances, by transelimination degradation, during treatment with enzymes. The degraded fragments may be lost during dialysis, if they are of sufficiently low molecular weight. We have checked this possibility with apple fibre and the results are discussed in Section 3.3.3.

3.3.2. Observations on the Method of Schweizer and Würsch

The procedure of Schweizer and Würsch is similar to that of Theander and Aman, but was developed independently. In this procedure also, the total DF is measured as the sum of the buffer-soluble and buffer-insoluble DF fractions. For details of the method see refs. 19 and 119. The basic steps involved in the method are shown in Fig. 3. The following aspects of the method should be noted: (i) the gelatinisation of the free starch is accomplished by autoclaving an aqueous suspension of the product at 130 °C for 20 min; (ii) the proteins (present in the AIR) are removed by digestion with pepsin before the enzymatic degradation of starch with glucoamylase; (iii) 0·5 M H_2SO_4 hydrolysis for 3 h at 100 °C is used for liberating the sugars from the non-cellulose polysaccharides, and the lignocellulose complex is determined gravimetrically; (iv) the neutral sugars are determined by GLC as their aldononitrile acetates and the uronic acids are estimated by a modification of the carbazole reaction.

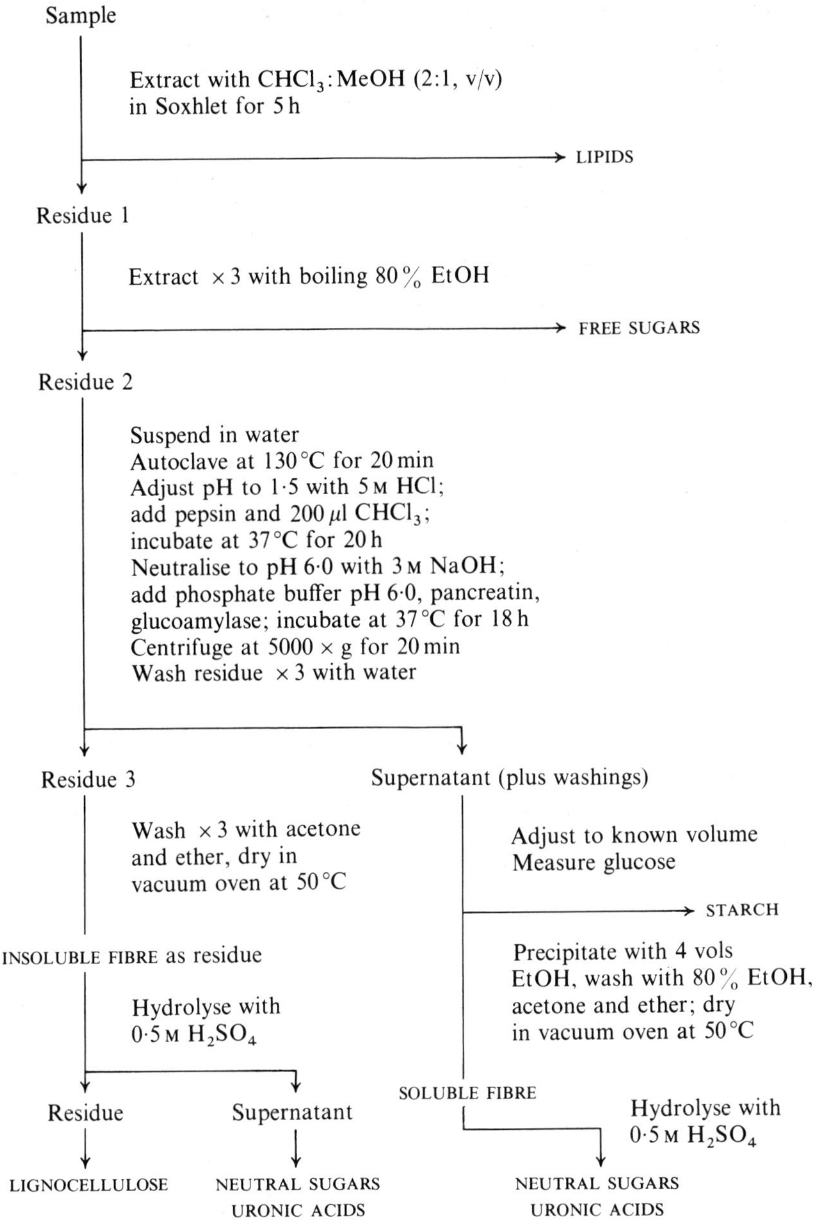

FIG. 3. Scheme of analysis (after Schweizer and Würsch[119]).

This method has been used to determine the DF content of a range of cereal and vegetable products and has been found to be satisfactory, particularly in view of the results given in refs. 118 and 119. The procedures used for removing proteins and starch appear to be very effective, and it is likely that the enzymatic degradation of proteins facilitates subsequent starch hydrolysis. This point needs special emphasis because the DF contents of wheat bran, rye flour (A), rye biscuit (A) and potato powder, as given in Table 3 of ref. 119, clearly show that except for potato powder the bulk of the starch can be degraded enzymatically after removal of proteins, even without the gelatinisation step. Although the authors state that under the experimental conditions used, pectins are not likely to be degraded, very little, if any, supporting evidence is given.[119] The results with commercial citrus pectins are not conclusive because they are partially modified and may not be degraded as readily as native pectins.

3.3.3. Observations on the Simplified Procedure of Selvendran and DuPont

We have developed two procedures for the analysis of dietary fibre. One is based on the isolation and analysis of cell wall material from plant tissues, which is suitable for preparing and analysing gram quantities of relatively pure DF,[23] and the other is based on the enzymatic removal of starch from the alcohol-insoluble residue, which is suitable for preparing and analysing milligram quantities of DF.[16] The first approach will be discussed in detail in Section 5. Here we shall consider only the relative merits of the second procedure, which is shown schematically in Fig. 4. Most of the steps involved in the development of the second procedure are based on methods developed for cell wall analysis. These are: (i) 'complete' disruption of tissue structure by wet ball-milling;[15,23] (ii) removal of starch enzymatically with a mixture of α-amylase and pullulanase;[16] originally salivary α-amylase was used but subsequent work has shown that pancreatic α-amylase is as effective and cheaper; (iii) the use of different acid hydrolysis conditions for the release of neutral sugars from the non-cellulosic polysaccharides and cellulose; (iv) determination of the neutral sugars by GLC of the alditol acetates;[33] and (v) determination of the uronic acids by the modified carbazole method.[33] Most of the recommendations have been incorporated into the method described in ref. 16.

However, since that publication we have had some queries on the following aspects of the method: (a) the need for wet ball-milling the alcohol-insoluble residues of all food products; (b) possible degradation of pectic substances (to dialysable fragments) during the gelatinisation step,

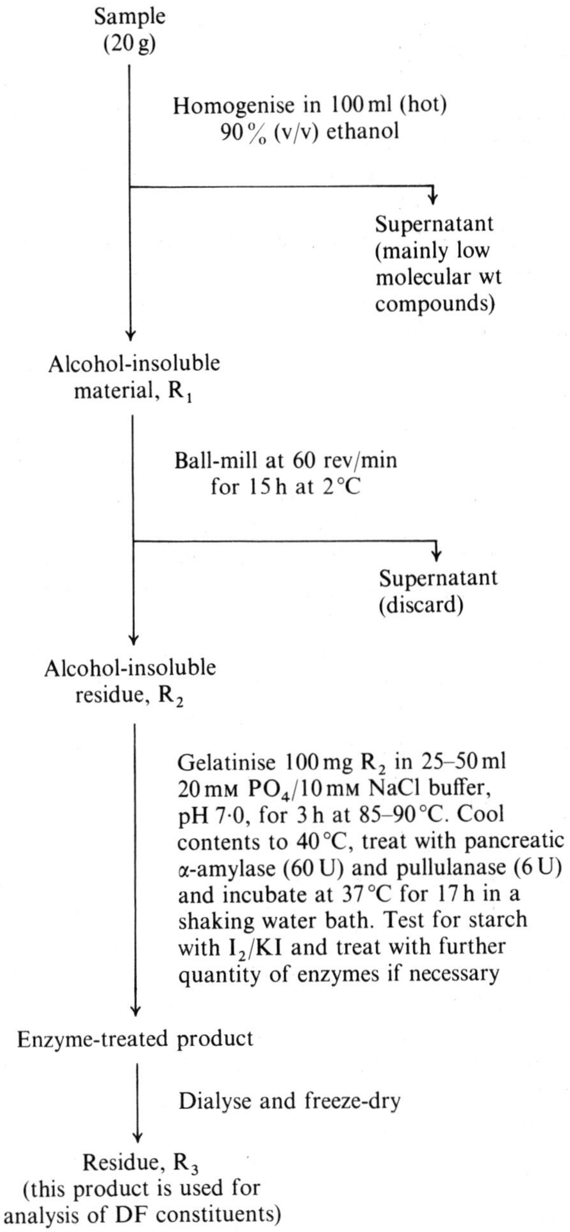

FIG. 4. Scheme of analysis (after Selvendran and DuPont[16]).

and the losses if any; (c) an estimate of the DF polymers solubilised by the buffer; (d) the occurrence of 'resistant starch' in some products; (e) complete release of uronic acid-containing polymers from the DF preparations on treatment with 72% H_2SO_4 and dilution to 1 M acid; (f) alternative methods for uronic acid analysis, where the interference from neutral sugars would be negligible; (g) comparison of DF values obtained by GLC and colorimetric techniques; (h) relative merits of the various methods available for lignin determination. In view of these queries, we have carried out several checks and the results of these investigations will be discussed because they are relevant to most of the methods that have been put forward for DF analysis.

(a) Wet Ball-Milling Step

Our work on the isolation of cell wall material from starch-rich tissues, particularly cereals, has shown that ball-milling the products in aqueous solvents is necessary for complete disruption of tissue structure.[15,23] In the case of alcohol-insoluble residues of processed foods, particularly from cereals, we have found that ball-milling the product in 90% aqueous ethanol facilitates the subsequent removal of starch enzymatically. For products which are not very rich in starch, fine-milling the AIR to ensure homogeneity of the sample, with a particle size of 0·5–1 mm, has been found to be adequate. Even with starch-rich products like potatoes, oats and rye biscuit, fine-milling alone has been found to be adequate sometimes, but the results are not always reproducible because the removal of starch is not always 'complete'. However, it was found that the hydrolysis of the purified DF preparations, particularly those containing lignified tissues, was more complete with the ball-milled samples; the amount of neutral sugars released was about 10% higher than those obtained with the fine-milled samples (Selvendran and DuPont, unpublished results). Theander and Aman, as well as Schweizer and Würsch, claim complete removal of starch enzymatically from starch-rich products, although they do not finely subdivide the alcohol-insoluble residues by wet ball-milling. This could be due to the fact that the first group of workers degrade the bulk of the starch with Termamyl α-amylase at a fairly high temperature (85 °C). However it should be noted that Termamyl α-amylase exhibits considerable glucanase activity at a lower temperature (40 °C) (Englyst, unpublished results). The second group of workers pretreat the AIR with pepsin to remove the bulk of the proteins and this treatment probably renders the starch, associated with proteins, more readily degradable enzymatically. Englyst has achieved complete removal of starch enzymatically from starch-rich products by avoiding treatment with alcohol and thus causing minimum modifications to the starch.

As the wet ball-milling procedure may degrade some of the cell wall polymers, suitable checks were carried out with potato starch and lysozyme. The results with these products[23,25] showed that the cell wall polymers are unlikely to be degraded under the conditions used in our studies. The following additional experiments with cell wall material from wheat bran and apples confirmed these results: the cell wall preparations were ball-milled with 90% ethanol for 20 h and the supernatants were concentrated and analysed for sugars after hydrolysis with 1 M H_2SO_4 for 2·5 h before and after dialysis. The amounts of carbohydrate polymers present in the supernatants before and after dialysis, expressed as a percentage of the dry weights of the CWM, were as follows: wheat bran 0·1% and 0·05%, and apples 0·5% and 0·05%. From these results it can be concluded that very little, if any, dietary fibre polymers are degraded and lost as a result of ball-milling the alcohol-insoluble material in 90% aq. ethanol.

(b) Possible Degradation of Pectic Substances During Gelatinisation

For gelatinising the starch, the alcohol-insoluble residues in 20 mM PO_4/10 mM NaCl buffer (pH 7·0) are heated at 85–90 °C;[16] the time of heating can vary from 1 to 3 h depending on the product analysed. The pH of the buffer should be ~7·0 because this is optimum for α-amylase from hog pancreas and human saliva.[120] During heating some of the pectins may be degraded by transelimination degradation to low molecular weight fragments, which may be lost during dialysis. To check this point, the following experiments were carried out with AIR of apples. This material was chosen because apple pectin has a high degree of esterification and has associated neutral sugar residues. One sample was gelatinised directly and the other was de-esterified by treatment with dilute NaOH (pH 11·75) at 0 °C for 2 h as described in ref. 121, neutralised to pH 7·0 with phosphoric acid, and then gelatinised. It was assumed that the de-esterified material would be less susceptible to transelimination degradation. The main steps involved in the analysis and the recovery of the various products are shown in Fig. 5 and the results of analysis are shown in Table 6. The total amount of carbohydrate material detected in the diffusates from the buffer-soluble fractions accounted for about 0·6% of the carbohydrate present in the starch-free AIR. From the combined results, that is the amounts of material recovered and the carbohydrate compositions of the buffer-soluble and buffer-insoluble residues and diffusates, it appears that very

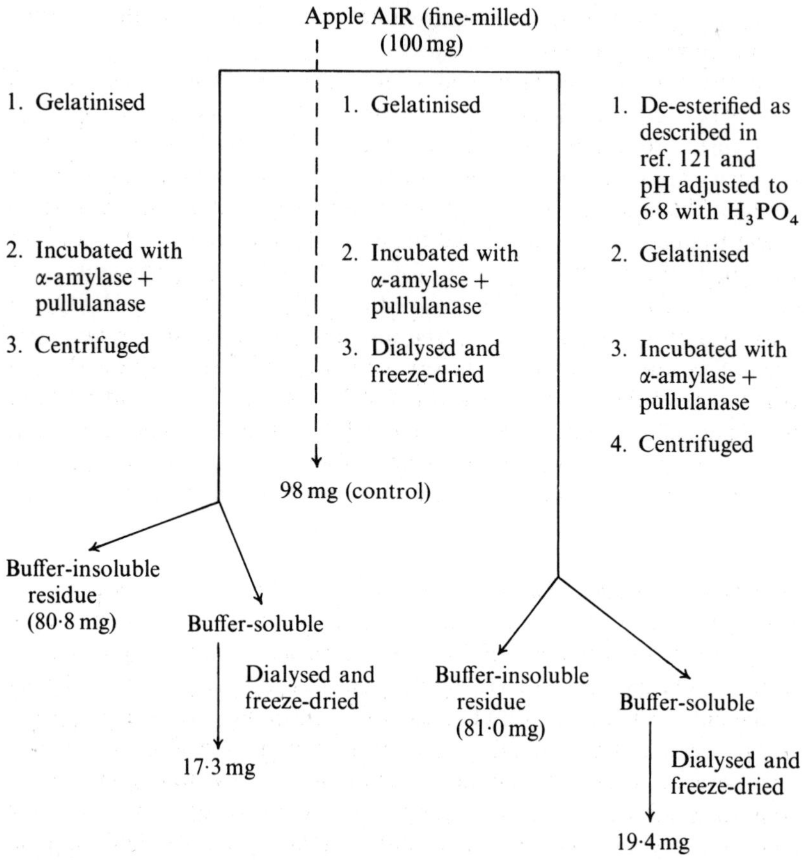

FIG. 5. Fractionation of AIR of apples with and without de-esterification.

little, if any, dietary fibre polymers are lost as a result of transelimination degradation. Probably some of the pectins are degraded, but they are not reduced to dialysable fragments.

(*c*) *Material Solubilised by Buffer*

We consider that there is very little advantage in separating the buffer-soluble DF polymers from the insoluble residue. The amount of material solubilised is dependent on the previous history of the sample, for example the dehydration conditions to which it has been subjected. In the case of cereal products, the amount of DF polymers solubilised is usually <10%

TABLE 6

CARBOHYDRATE COMPOSITIONS OF BUFFER-INSOLUBLE AND BUFFER-SOLUBLE FRACTIONS FROM AIR OF APPLES[a]

	Air (not de-esterified)		*AIR (not de-esterified)*	*AIR (de-esterified)*	
	Buffer-sol.	*Buffer-insol.*	*Buffer-sol- + buffer-insol. (control)*	*Buffer-sol.*	*Buffer-insol.*
Rha	13·3	26·6	23·0	10·3	25·0
Ara	94·1	128·4	128·3	95·8	127·6
Xyl	5·3	64·4	48·7	8·5	62·5
Man	2·2	23·5	18·5	1·9	22·8
Gal	38·6	64·8	57·1	32·7	63·0
Glc	27·5	383·3	300·2	28·9	365·0
	(13·8)	(14·7)	(14·1)	(10·5)	(15·5)
Uronic	657·0	233·7	331·2	600·0	250·0

[a] Neutral sugars released on Saemen hydrolysis; values are expressed as μg anhydro-sugar/mg dry preparation. The values within parentheses represent the amount of glucose released on 1 M H_2SO_4 hydrolysis.

of the total DF, whereas in the case of vegetable and fruit products as much as 10–15% of the total DF could be solubilised; the solubilised material is mainly of pectic origin (Selvendran and DuPont, unpublished results).

(d) The Occurrence of 'Resistant Starch'

In our studies with starch-rich tissues (e.g. fresh potatoes and oats) it has been found that although about 98% of the starch could be degraded enzymatically, about 1–2% of the starch (apparently) resisted enzymatic hydrolysis; this is inferred from the 1 M acid hydrolysis values for glucose (Table 5, col. 8). The bulk of this 'resistant starch' is not dialysable, but is soluble in 80% (v/v) aq. alcohol, and remains in the supernatant when the enzyme digest is made 80% (v/v) with respect to alcohol. The occurrence of the resistant starch would be apparent in the method of analysis for DF which we have proposed (a procedure which does not involve alcohol precipitation of the enzymatic digest), but would not be apparent in the method proposed by Englyst *et al.* (which is a procedure that involves alcohol precipitation). The nature of the 'resistant starch' is not clear, and does not seem to arise from insufficient starch-degrading enzymes. If starch is a proteoglycan, it is possible that some of the glucose residues linked to protein are not enzymatically degradable but are acid hydrolysable.

(e), (f) Determination of the Uronic Acid Content

In our earlier papers[16,33] the determination of uronic acid-containing polymers in the dietary fibre preparations by a modified carbazole method was described. This method involves first dispersing the dietary fibre preparation in 72% H_2SO_4 for 3 h at 20 °C, followed by dilution to 1 M acid, filtration through glass fibre filter paper, and then taking an aliquot for analysis. Since the neutral sugars gave a small response (5–15%) to the carbazole reaction, the estimate of uronic acid was corrected for neutral sugars liberated on Saeman hydrolysis of the preparation.

We have now improved on this method in two ways: (i) The material is dispersed in 72% H_2SO_4 for 3 h at 20 °C diluted for 1 M acid, and then heated for 1 h at 100 °C before cooling and filtration. The dissolution and hydrolysis of the uronic acid-containing polymers appears to be more complete compared with the earlier method, and the values for uronic acid are about 10% higher for most preparations. (ii) The *m*-phenyl phenol method for the determination of uronic acid[122] is preferred over the modified carbazole method because the interference from neutral sugars is negligible and no correction has to be made for them (Selvendran and DuPont, unpublished results; see also ref. 123). However, the uronic acid values from fibre preparations from apples, potatoes, wheat bran and oats by both methods are highly comparable.

(g) Comparison of Sugars in Fibre Preparations by GLC and Colorimetric Methods

A comparison of the sugars estimated by colorimetric and GLC methods in some fibre preparations is shown in Table 7. The sugars were liberated by Saeman hydrolysis of the fibre preparations and aliquots were taken for analysis. The pentoses and hexoses were estimated by the phloroglucinol[124] and anthrone[125] methods respectively, and the neutral sugars were estimated by GLC as the alditol acetates.[33] The uronic acid content was determined by the *m*-phenyl phenol method discussed above.

From Table 7 the following conclusions may be drawn: (1) For wheat bran, there is quite good agreement between estimates of pentoses by GLC and colorimetry, but colorimetry overestimates the hexoses. Since the uronic acid content of the preparation is low, it hardly interferes with pentose estimation by colorimetry. Further, hexoses do not interfere with pentose estimation unless they are present in excessive amounts. Therefore, one would expect good agreement by colorimetry and GLC. However, since pentoses give an appreciable response to the anthrone method[62] (also Selvendran and DuPont, unpublished results), the hexose values would be considerably higher. With the anthrone reagent, pentoses give a brownish-yellow colour as compared with the bluish-green colour given by the hexoses. (2) For rye biscuit, the estimate of pentoses by colorimetry is appreciably lower than that obtained by GLC and the reason for this is not clear. The hexose value by colorimetry, as expected, is higher than that obtained by GLC. (3) For both runner beans and apple preparations, the pentose values by colorimetry are considerably higher than those obtained

TABLE 7

COMPARISON OF GLC AND COLORIMETRIC METHODS FOR ESTIMATING SUGARS IN FIBRE PREPARATIONS

(Values are μg anhydro-sugar/mg dry preparation)

Sugars[a]	*Wheat bran*		*Rye biscuit*		*Runner bean*		*Apples*	
	GLC	*C*	*GLC*	*C*	*GLC*	*C*	*GLC*	*C*
Rha	1·9		2·1		11·3		19·4	
Ara	129·2	367·1* (359·3)	95·1	199·9* (232·1)	25·2	149·6* (93·0)	119·3	233·7* (167)
Xyl	230·1		137·0		67·8		47·7	
Man	5·6		19·9		31·0		18·5	
Gal	7·9	390** (201·2)	9·0	433** (290·9)	68·0	575·5** (415·7)	54·7	553·4** (341·4)
Glc	187·7		262·0		316·7		268·2	
Uronic	[35·2]		[21·6]		[140·0]		[277·2]	

[a] The sugars released on Saeman hydrolysis are given.
Under column C, * gives colorimetric values for pentoses, and ** gives colorimetric values for hexoses; the values within parentheses give the total amount of pentoses and hexoses estimated by GLC. The values for uronic acid were obtained by a colorimetric method. The preparations from wheat bran, rye biscuit and apples were obtained by treating the AIR with α-amylase + pullulanase, whereas in the case of runner beans the CWM was used.

by GLC. This is probably due to interference from galacturonic acid of pectins. Galacturonic acid gives about 25% of the response of xylose by the phloroglucinol method (Selvendran and DuPont, unpublished results). The hexose values by colorimetry are also considerably higher than those obtained by GLC, and these discrepancies can also be traced to interference, mainly from galacturonic acid, although the relatively low levels of pentoses would also interfere. Thus it would appear that the estimation of sugars by colorimetric methods is not very reliable.

(*h*) *Lignin Determination*
The various methods available for the estimation of lignin in DF preparations and their limitations have been discussed in Section 2.3. We find it difficult to recommend any particular method, but consider that the Klason lignin value gives a fair estimate of the lignin content of the preparation, provided it is free of the bulk of starch and intracellular proteins. When very small amounts of the material are available, the acetyl bromide method is to be preferred. However, each worker has to determine the relative merits of the methods on some well-defined samples.

3.3.4. OBSERVATIONS ON THE PROCEDURE OF ENGLYST *et al.* AND CHEN AND ANDERSON

(*a*) *Procedure of Englyst* et al.
The procedure of Englyst as described in ref. 26 measures the total DF as the sum of the buffer-soluble and buffer-insoluble dietary fibre fractions. In this method, the (freeze-) dried food sample is first gelatinised by heating with acetate buffer at pH 5 for 5 h at 90 °C. The suspension is then cooled and treated with amyloglucosidase for 16 h at 45 °C, and centrifuged to separate the water-soluble and water-insoluble DF fractions. From the supernatant fluid the water-soluble non-cellulosic polysaccharides are obtained by precipitation with alcohol, and determined after hydrolysis with 0·5 M H_2SO_4 for 2·5 h. Similarly, the water-insoluble fraction is hydrolysed with 0·5 M H_2SO_4 for 2·5 h to release the sugars from the non-cellulosic polysaccharides. The residue from above is solubilised in 72 % H_2SO_4 for 24 h, diluted and filtered, and a measure of cellulose obtained by colorimetric estimation of the hexose content of the filtrate. Using the above procedure, Englyst has determined the DF content of several food products, and the dietary fibre values obtained by his method compare favourably with those obtained by the procedure of Theander and Aman.[118]

In a subsequent paper, Englyst *et al.*[24] claim that they have improved on the above procedure in the following ways: (1) by using more effective and more specific enzyme preparations for the hydrolysis of starch; (2) by using a more informative separation of non-starchy polysaccharides; and (3) by obtaining an estimate of the modified starch, which is resistant to α-amylase digestion. A scheme outlining the main points of their improved method is shown in Fig. 6.

The following points should be noted with respect to the improved procedure: (1) The gelatinisation time is reduced from 5 h to 1 h, but the reservation is made that an extended period (up to 3 h) may be necessary for

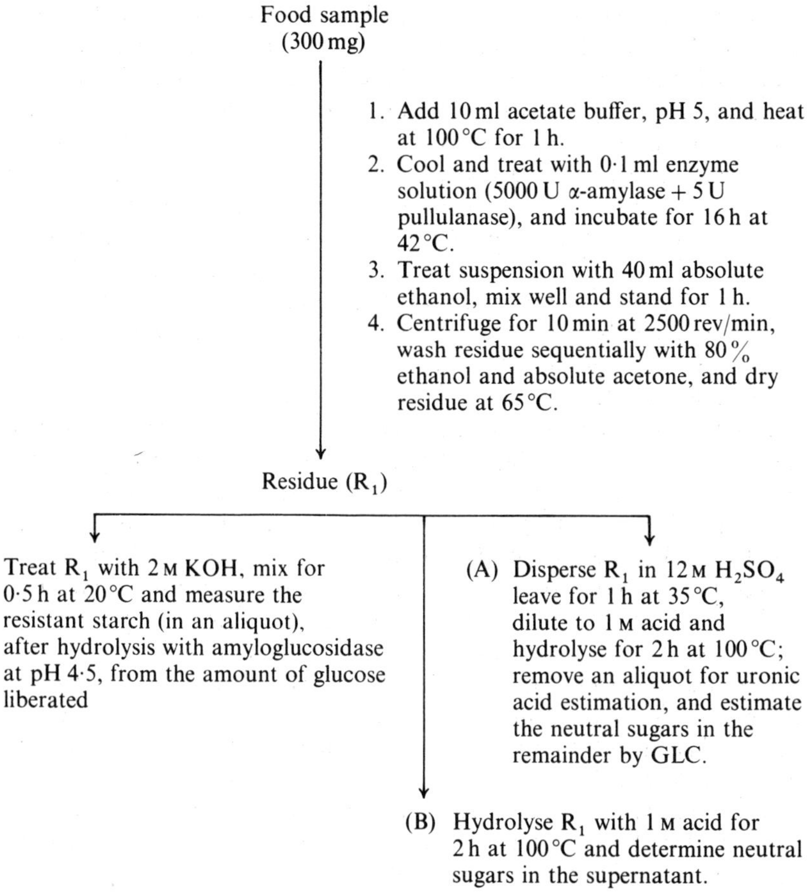

FIG. 6. Scheme of analysis (after Englyst *et al.*[24])

some starch-rich products. This finding compares well with that of Selvendran and DuPont.[16] (2) A mixture of hog pancreatic α-amylase and pullulanase is used instead of amyloglucosidase for the degradation of starch. This is because the first two enzymes were shown to be free of β-glucanase activity. The use of mammalian α-amylase and pullulanase for the 'complete' degradation of starch from cereal and vegetable products has been described in some of our earlier papers;[15,16] see also ref. 126. A disturbing feature about the use of pancreatic α-amylase in the procedure of Englyst *et al.*[24] is that a very large excess of enzyme is recommended—

5000 units for 200 mg of food product. We have found that for 200 mg of starch-rich product 80 units of enzyme would be sufficient;[16] in subsequent work we have confirmed this but would recommend 120 units of pancreatic α-amylase to allow a safety margin. Thus the amount of α-amylase recommended by Englyst *et al.* is >40 times the actual amount required. An explanation for this discrepancy could be traced to the pH of the enzyme medium. Englyst *et al.* incubate in acetate buffer at pH 5·2, whereas it is stated in the literature that mammalian α-amylase is unstable at a pH less than 6·9–7·0;[120] in our studies we have incubated at pH 7·0. (3) The differential acid hydrolysis treatments recommended give an estimate of non-cellulosic polysaccharides and cellulose. We have described at length the relative merits of the 1 M H_2SO_4 and Saeman hydrolysis conditions to obtain a measure of non-cellulosic polysaccharides and cellulose,[16,33,127] and the reader is advised to consult these papers for fuller details. (4) The use of 2 M KOH to solubilise amylase-resistant starch and its subsequent estimation (from the amount of glucose released) after hydrolysis with amyloglucosidase (pH 4·5, 1 h at 65 °C) is particularly noteworthy. The incubation is carried out at 65 °C to inactivate β-glucanase, which may be present as a contaminant in some commercial amyloglucosidase preparations. In order to overcome the problems associated with β-glucanase activity, we used α-amylase and pullulanase to degrade the modified starch solubilised by alkali (4 M KOH–10 mM $NaBH_4$/2 h/20 °C, under argon). After dialysis, and hydrolysis of the dialysate, we found that we could measure the modified starch content of cornflakes satisfactorily as glucose either enzymatically or by GLC.

The procedure of Englyst *et al.* has been used to determine the DF content of a range of cereal and vegetable products, and the results are particularly valuable in the dietary fibre context.

(b) Procedure of Chen and Anderson

The method of Chen and Anderson[128] is similar to the procedure described above and involves six steps: (1) preparation of the 85 % methanol/ether-insoluble residue (as methanol is a harmful solvent, we recommend the use of ethanol); (2) gelatinisation in hot water for 10 min, followed by incubation with bacterial α-amylase, pH 6·9, at 25 °C overnight, to remove starch; (3) precipitation of the enzyme digest with 80 % ethanol; (4) extraction with hot water for 20 min to effect partial fractionation of water-soluble and water-insoluble DF fractions; (5) analysis of the fractions for sugars after hydrolysis with 1 N H_2SO_4 (2·5 h) or 2 N TFA (1 h); and (6) analysis of 1 N H_2SO_4-insoluble residue for cellulose and lignin.

This method has been applied to determine the DF content of a range of food products. However, no comment has been made on the possible presence of hemicellulase activity in the bacterial α-amylase preparation. Most of the general comments made in the earlier sections are equally applicable to this method.

3.4. Enzymatic Determination of Dietary Fibre

The enzymatic method was proposed by Hellendoorn *et al.*[129] and depends on the *in vitro* digestion of food products with pepsin and pancreatin. The novel feature of this method is that it is based on the use of alimentary digestive enzymes. Although the original method measured only the insoluble indigestible residue, in a subsequent paper[130] the measurement of soluble pectins is recommended. Application of the above method may not result in the 'complete' degradation of starch and proteins, due to incomplete disruption of tissue structure and other factors. However, Hellendoorn claims that his method gives a better indication of non-digestible residue because it approximates to the physiological conditions existing in humans. Although there is an element of truth in his reasoning, for the purpose of this discussion we shall restrict the definition of dietary fibre to that given in the introduction to this chapter.

An improved version of the Hellendoorn type of analysis was used by Honig and Rackis[131] for the determination of the total pepsin/pancreatin-indigestible content of soybean products, wheat bran and corn bran. The modified procedure is illustrated in Fig. 7; both the soluble and insoluble indigestible fractions were isolated to determine the total indigestible residue content of the products. In this paper, the results of dietary fibre analysis of the various products are compared with some of the findings of Hellendoorn *et al.*, and thus give an indication of the water-soluble indigestible residues which are not measured by Hellendoorn *et al.* The carbohydrate analyses of the component sugars in the soluble residues appear to suggest that starch is probably not completely degraded by pancreatin.

Asp and Johansson[132a] have improved on the Hellendoorn type of analysis by measuring the soluble and insoluble dietary fibre components, and by including a gelatinisation step prior to digestion with pepsin and pancreatin. Gelatinisation appears to result in more complete enzymatic degradation of the starch. The improved method which they have developed is shown schematically in Fig. 8. It should be noted that in this method the dietary fibre is measured gravimetrically. Application of the method to determine the DF content of a range of food products is

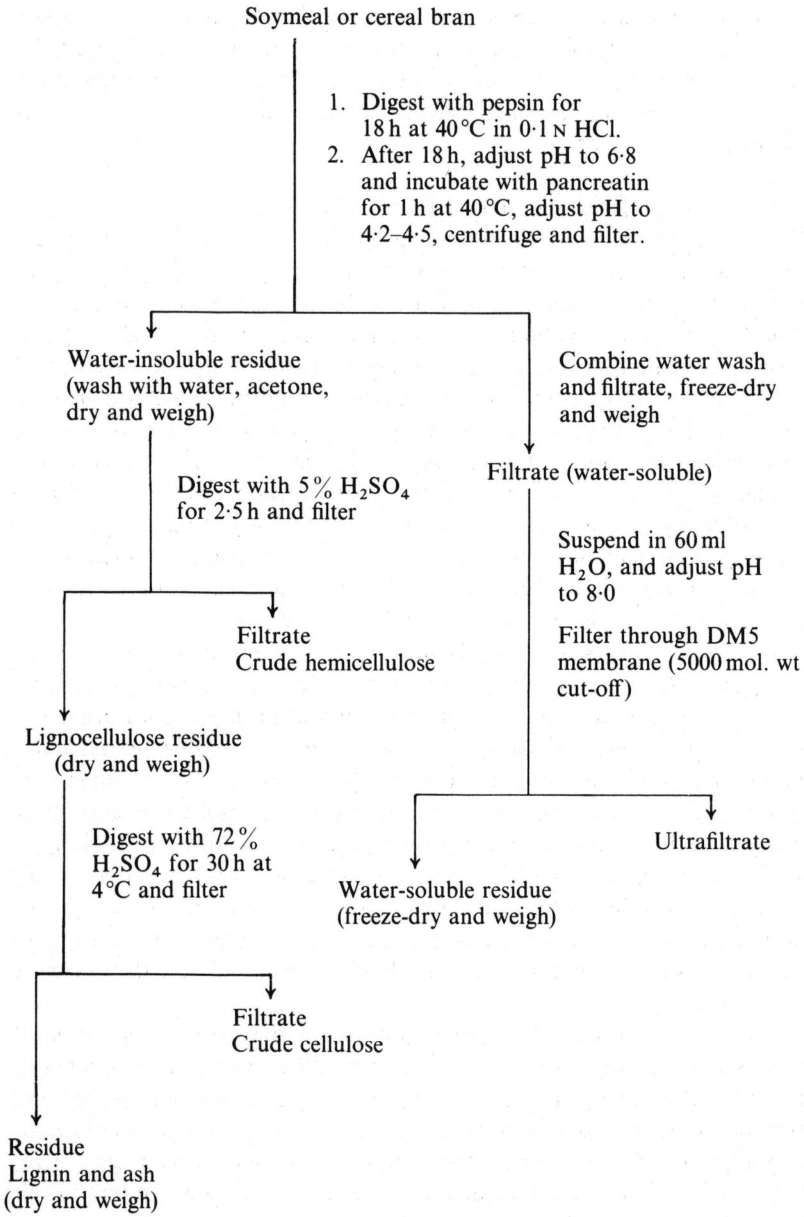

FIG. 7. Scheme of analysis (after Honig and Rackis[131]).

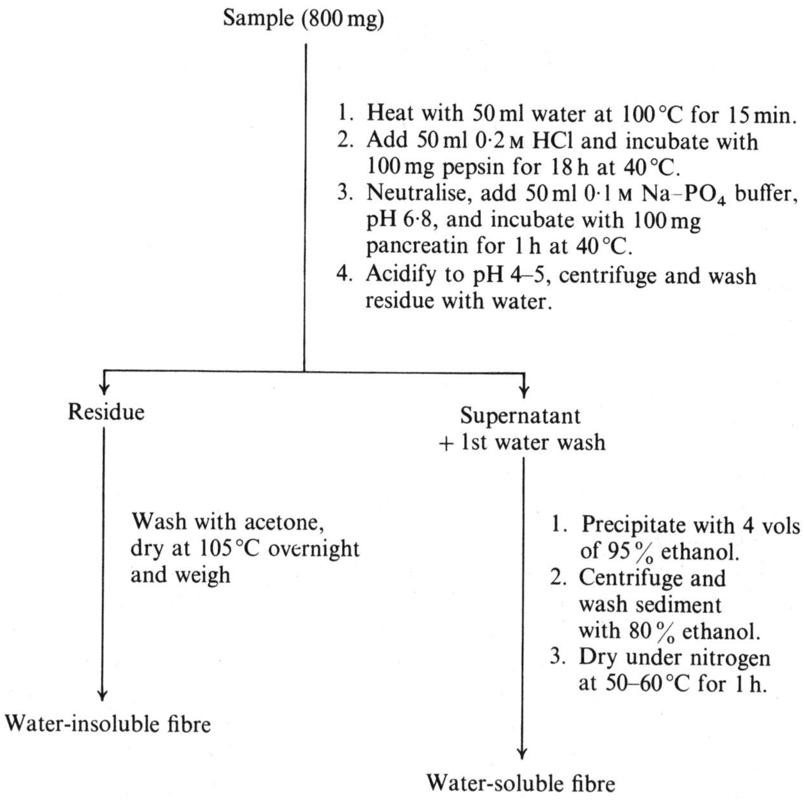

FIG. 8. Scheme of analysis (after Asp and Johansson[132a]).

discussed in ref. 132a, and it appears that the values obtained for DF content of a number of EEC samples compare favourably with those obtained by Theander and Aman.[118]

After this chapter was submitted, Asp *et al.*[132b] published an improved and rapid method for the determination of insoluble and soluble dietary fibre.

3.5. A Note on the Determination of DF Content of Food Additives

Although food additives such as guar gum, Ispaghula husk, xanthan gum, carrageenan, alginates, etc., are included under the (broader) definition of the term DF, very few workers have commented on the estimation of these compounds. A large proportion of these compounds are of cell wall origin

TABLE 8
ANALYSIS OF SOME FOOD ADDITIVES
(Values are μg anhydro-sugar/mg dry preparation)

Sugar	*Ispaghula*	*Xanthan gum*	*Carrageenan*	*Alginate*	*Guar gum*
Rha	33·9	1·9	3·1	n.d.	6·0
Ara	237·0	3·4	3·2	n.d.	21·5
Xyl	620·5	0·7	2·1	n.d.	3·9
Man	19·9	163·1	—	n.d.	562·0
Gal	39·3	—	277·1	n.d.	335·4
Glc	9·4	264·0	19·1	n.d.	24·9
Uronic acid	n.d.	200·8[a]	49·3	766·4	46·9

[a] UA was calculated using galacturonic acid as standard.
n.d., not determined.

(but from very diverse types of plants) and the nature of the sugars (and sugar derivatives) present in some of them poses some analytical problems, which call for a comment.

The carbohydrate composition and DF content of the food additives listed above are shown in Table 8. Guar gum and Ispaghula husk contain mainly galactomannans and arabinoxylans respectively, and the constituent sugars are quantitatively released on 1 M H_2SO_4 hydrolysis. The relatively small amount of uronic acid(s) present in Ispaghula husk arabinoxylans does not give rise to any special problems. Xanthan gum contains glucose, mannose and glucuronic acid in the ratio 2:2:1,[133,134] and the mannose residues linked to glucuronic acid (50% of mannose) are only slightly hydrolysed (Table 8, col. 2); this is due to the stability of the aldobiouronic acid (glucuronic acid (1→2)-mannose). Carrageenan contains galactose sulphate and 3,6-anhydrogalactose residues, and the amount of galactose released on acid hydrolysis is only about 30% of the expected value. Alginate contains mannuronic and guluronic acid residues, which give about 75% of the response of galacturonic acid by the Blumenkrantz method,[122] and this is reflected in its uronic acid content. These observations should be borne in mind when the contribution of food additives to DF content is being assessed.

4. GENERAL RECOMMENDATIONS FOR DIETARY FIBRE ANALYSIS

From the above survey we would like to make the following recommendations for dietary fibre analysis:

1. Preparation of sample. It is debatable whether the freeze-dried sample (FDS) or the alcohol-insoluble residue (AIR) should be used as the starting material. For products which are rich in lipids there is no alternative because it is necessary to use alcohol/benzene or alcohol/chloroform extraction. With the AIR of most products, provided suitable precautions are taken, the enzymatic degradation of precipitated starch and proteins can be made to go to 'completion'. AIR is convenient to prepare and is preferred for products which are rich in polyphenols because the bulk of the polyphenols can be removed with alcohol. The presence of polyphenols and other low molecular weight enzyme inhibitors could cause problems with freeze-dried samples of some foods.

The AIR or freeze-dried sample must first be milled to pass through a 0·5 mm mesh. Although we have recommended ball-milling the AIR in 90 % ethanol for 10 h at 2 °C to disrupt the tissue structure completely, we now think that this step may be unnecessary for some foods. However, hydrolysis appears to go to 'completion' with DF preparations from the ball-milled samples.

2. Gelatinisation of the AIR or freeze-dried sample in phosphate buffer, at pH 7·0, at 85 °C for 1–3 h is recommended; 1 h would suffice for most foods.

3. A mixture of pancreatic α-amylase + pullulanase is recommended for degrading starch. For 50 mg AIR of starch-rich product 20 U of α-amylase and 3 U pullulanase would suffice (but 30 U α-amylase is recommended), and it is important that the incubation medium is maintained at pH 7·0. The time of incubation should be ~18 h at 37 °C. After this, a small amount of the sample should be removed and tested for residual starch with I_2/KI. If the residue stains blue with iodine, then the sample should be treated with enzyme mixture for an additional 7 h. If the suspension still gives a positive test with I_2/KI, it may be safe to assume that the product contains some modified starch. However, it should be noted that amylase-resistant starch may not always give a positive iodine test.

4. The buffer-soluble and buffer-insoluble DF fractions may be obtained by either freeze-drying the dialysed suspension from above, or by precipitating with alcohol (to 90 % alcohol concentration) and washing the precipitate by centrifugation with absolute ethanol and ether. If the removal of precipitated proteins is desired, then drying may be omitted. For the enzymatic degradation of proteins we have found that incubation with pronase is quite effective (for details see ref. 115), but several workers have obtained satisfactory results with pepsin.[119,131,132] The starch- and protein-free AIR would give a good estimate of the total DF provided a correction for residual proteins and ash is made.

5. A measure of the amylase-resistant starch from processed foods can be obtained by extracting the enzyme-treated residue with 2 M KOH or 4 M KOH, containing 10 mM $NaBH_4$, preferably for 2 h at 20 °C to solubilise the resistant starch. The solubilised starch can be determined from the amount of glucose released either on treatment with amyloglucosidase at 65 °C or, preferably, after treatment with α-amylase and pullulanase, followed by hydrolysis of the dialysate.

6. For convenience, the component DF polymers could be classified as non-cellulosic polysaccharides (NCP) excluding polyuronides, cellulose, polyuronides and lignin. Although the bulk of the sugars from the NCP (and about 5–10 % of the cellulose) can be hydrolysed with 1 M H_2SO_4 at 100 °C in 2·5 h, the values obtained for sugars from non-cellulosic polysaccharides on Saeman hydrolysis are slightly higher, particularly for xylose and mannose, and we consider that the Saeman hydrolysis gives the better estimate of the sugars from NCP. An estimate of 90–95 % of the glucose released from cellulose can be obtained from the difference in glucose values obtained by Saeman hydrolysis and 1 M H_2SO_4 hydrolysis. The sugars released by hydrolysis can be determined as the alditol acetates by GLC.

For determining the total uronic acid content of DF, the modified carbazole method or the *m*-phenyl phenol method is recommended as it is more convenient than the decarboxylation method. The relative merits of these methods are briefly discussed in Section 2.2.

An accurate estimate of the lignin content of the dietary fibre preparation is difficult to obtain. The Klason lignin value of the starch- and protein-free AIR should give a fairly good estimate of the lignin content, but if only very small amounts of material are available the acetyl bromide lignin value may be obtained. The permanganate oxidation method (gravimetric) gives a good estimate of the lignin content of the acid detergent residue, but not of the purified DF preparation.

5. PREPARATION AND ANALYSIS OF GRAM QUANTITIES OF DIETARY FIBRE

In this section we shall outline some of the improved methods which have been developed in our laboratory for the preparation and analysis of gram quantities of relatively pure dietary fibre from fresh (and processed) foods. Clearly, the most suitable procedure for the preparation of relatively pure DF would be one which causes minimal co-precipitation effects, and which, at the same time, ensures that the bulk of the intracellular compounds and only small amounts of the DF constituents are solubilised by the treatments used. The success of our methods depends on the 'complete' disruption of tissue structure by wet ball-milling and on the use of simple substances which have a high affinity for cytoplasmic molecules.

5.1. Preparation of Cell Wall Material from Fresh Tissues

A brief outline of the method is as follows. The plant material is blended with an Ultraturrax with 1% aq. sodium deoxycholate (SDC) (or 1% sodium lauryl sulphate, SLS) containing 5 mM sodium metabisulphite and the triturated material is wet ball-milled for 15 h at 2 °C. This gives a homogeneous product which can readily be separated by centrifugation. It should be noted that although aq. Na deoxycholate was recommended in some of our earlier papers[11,15,25] for solubilising intracellular proteins, we now prefer the use of aq. Na lauryl sulphate solution. The supernatant is decanted and retained for analysis of the soluble polysaccharides. The residue is washed with distilled water on the centrifuge. The wet residue is then extracted with phenol:acetic acid:water (PAW) (2:1:1, w/v/v); a short treatment in a blender gives a uniform suspension. The residue obtained by centrifugation (12 000 rev/min for 20 min) is washed with distilled water and rendered free of starch by extraction with 90% aqueous dimethylsulphoxide (DMSO) in an ultrasonic bath. After a few water washes the residue is freeze-dried. The relative merits of the de-starching procedures are discussed in refs. 15 and 25. The method with the relative amounts of material and extractants used is shown schematically in Fig. 9. For full details of the method, the reader is recommended to consult the above references, particularly ref. 15.

In the above method, the following points should be noted: (1) Wet ball-milling the triturated material is necessary to disintegrate the cell structure and render the contents accessible to solvents. The relative merits of the wet ball-milling step have been discussed earlier. (2) A small amount of the cold water-soluble pectic substances (in the case of vegetables and fruits) and some β-glucans and arabinoxylans (from cereals) would be solubilised by aq. SDC or aq. SLS. However, these polysaccharides could be isolated from the extracts and freed from contaminating starch and proteins by enzymatic methods. The amount of cell wall polysaccharides solubilised by aq. SDC when expressed as a percentage of the carbohydrate content of the purified cell wall material (CWM) is as follows: potatoes 4·1, apples 6·1, runner beans 8·5, wheat bran 1·0, and oats 13·7. These values refer to analysis of fresh tissues. (3) Phenol:acetic acid:water treatment desorbs the

Sample (10 g, dry wt)

Homogenise in 200 ml
1% aq. SDC or SLS.
Ball-mill at 2 °C for
15 h. Centrifuge

Residue, R_1

Supernatant (mainly intracellular compounds, some DF polymers)

Extract with 100 ml
PAW at 20 °C

Residue, R_2

Supernatant (residual proteins, adsorbed SDC, starch, etc.)

Extract with 200 ml
90% aq. DMSO at 20 °C

Supernatant (mainly starch, some glucans?)

Dietary fibre
only if fresh food

Treat 100 mg in 50 ml 20 mM PO_4/10 mM NaCl buffer, pH 7·0, containing α-amylase (40 U) and pullulanase (6 U), for 17 h at 37 °C

DF from starch-rich processed food

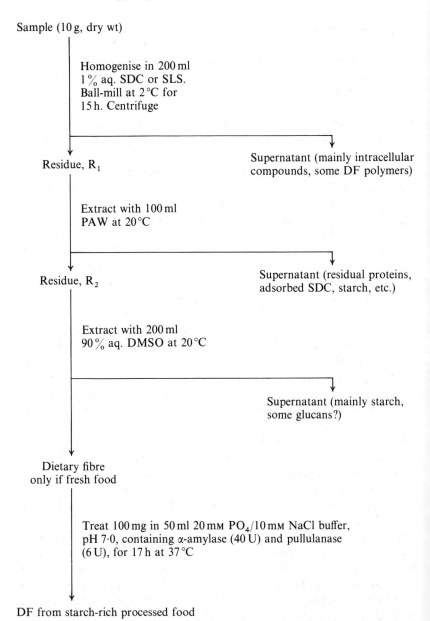

FIG. 9. Scheme for the preparation of DF from fresh and processed starch-rich products.

residual intracellular proteins, some starch, adsorbed SDC or SLS, lipids and pigments from the SDC (or SLS) residue, Very small amounts (<1 %) of cell wall polysaccharides are solubilised during the PAW treatment; however, a small proportion of starch is solubilised from some starch-rich products. (4) With all the starch-rich tissues (potatoes, oats and rye flour) which have been examined, 90 % aq. DMSO quantitatively solubilised the remaining starch; this was more than 90 % of the total starch in the tissue. Evidence to show that aq. DMSO does not solubilise more than ~2 % of CWM of potatoes has been reported.[25] In the case of oats (and rye flour) it appears that some of the β-glucans are also solubilised.[15] However, it is not clear whether the solubilised β-glucans are of cell wall or intracellular origin. (5) It is known that absolute DMSO solubilises some hemicelluloses (acetylated xylans) from the holocellulose (delignified CWM) of woody tissues[135] and a small amount (~3·5 %) of CWM from grass preparations ball-milled for 24 h.[136] We have found (unpublished results) that absolute DMSO, which is more effective in solubilising some polysaccharides than 90 % DMSO, solubilised <3 % of xylans from the CWM of parchment layers of mature runner bean pods at 25 °C, and that this amount increased to ~20 % with the holocellulose (i.e. the delignified material). (6) The SDC (or SLS)/PAW/aq. DMSO-treated preparations from starch-rich processed foods (e.g. potato powder and rye biscuit) had small amounts of starch associated with them; this starch, which may have been retained by hydrogen bonding, can be removed by treatment with pancreatic α-amylase. Therefore, the recommended method for the preparation of appreciable amounts of DF from processed starch-rich products would require two steps: (a) treatment with SDC (or SLS)/PAW/aq. DMSO to remove the bulk, i.e. more than 95 % of the starch and other intracellular compounds; and (b) digestion with pancreatic α-amylase to remove the residual starch. For processed products which have appreciable amounts of co-precipitated proteins, a preliminary treatment with pronase to remove the precipitated proteins is recommended (for details see ref. 115).

Alternative methods for the preparation of CWM from starch-rich tissues are given in refs. 137–140.

5.2. Sequential Extraction of Cell Wall Material

To obtain an indication of the types of carbohydrate polymers constituting the cell wall complex, chemical or enzymatic methods could be used. The latter type of methodology requires highly purified enzymes and has been used successfully for fractionating cell walls from sycamore suspension callus.[141–145] In this section we shall consider, briefly, only the chemical fractionation techniques which we have used with some success on CWM from a range of food products: potatoes,[25,116] runner beans,[12,78,79,146] cabbage,[115,147] apples and wheat bran.[52,148]

5.2.1. CWM FROM TISSUES CONTAINING APPRECIABLE AMOUNTS OF LIGNIN

If the tissue is relatively rich in lignin, then the extraction scheme outlined in Fig. 10 is recommended. For convenience the relative amounts of materials and extractants used are given (see also ref. 40). The method involves depectination of the CWM by sequential extraction with hot water (80 °C) at pH 5·0, and hot ammonium oxalate (80 °C) at pH 5·0, followed by delignification with sodium chlorite/acetic acid at 70 °C to give the holocellulose. From this product the hemicelluloses are extracted with 1 M and 4 M potassium hydroxide (KOH) containing 10 mM sodium borohyd-

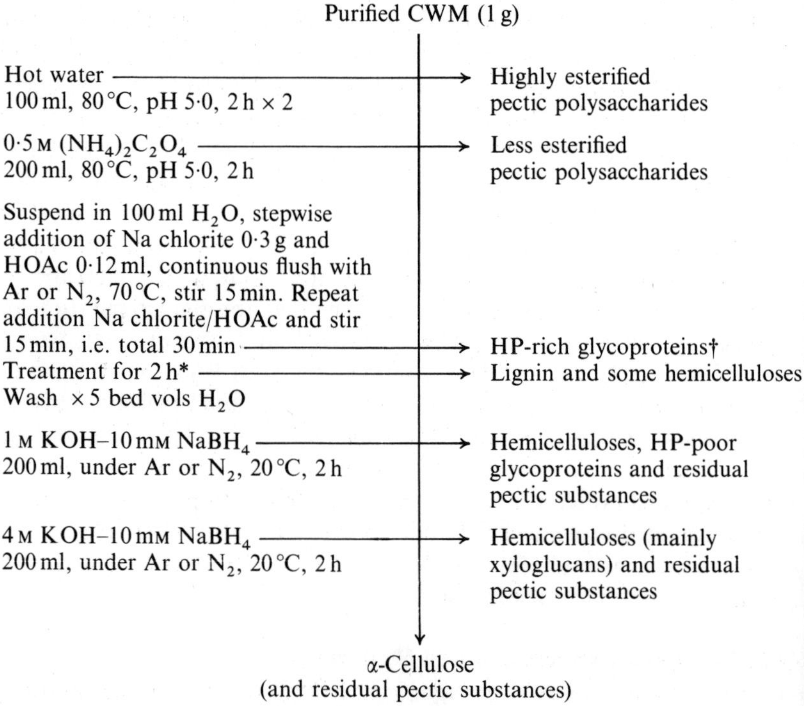

* For most tissues, 2 h appears to be adequate.

† The bulk of the hydroxyproline-rich proteins from the CWM of parenchymatous tissues are solubilised after 30 min.

FIG. 10. Scheme for the fractionation of purified CWM.

ride at 20 °C to leave a residue of α-cellulose. We prefer sequential extraction of the holocellulose (or depectinated material from tissues containing little or no lignin) with 1 M or 4 M KOH, because some measure of fractionation of the hemicelluloses can be effected. The hemicelluloses extracted with alkali are partially precipitated on neutralisation or acidification to pH 5·0 with acetic acid, and additional material can be precipitated by subsequent addition of an excess of acetone or alcohol.[149] Under these conditions, a small proportion of the hemicellulosic material remains in solution, and is usually not recovered. However, this material can be isolated by concentrating the supernatant, dialysing the concentrate and freeze-drying the dialysed material.

The conditions of extraction are chosen to enable partial fractionation of the main types of pectins and hemicelluloses, and to minimise (1) the breakdown of pectins by transelimination degradation, and (2) the hydrolysis of furanosidic linkages. In general, the hot-water-soluble pectins have a higher degree of esterification compared with the oxalate-soluble ones. Further, the 4 M KOH-soluble fraction has been found to be generally much richer in xyloglucans.[116,146] In the case of tissues containing a relatively high proportion of hydroxyproline (HP)-rich cell wall glycoproteins, the bulk of the HP-rich glycoproteins are solubilised by chlorite/HOAc during the delignification stage.[78,79] This property has been used for isolating 'modified hydroxyproline-rich glycoproteins' from the depectinated CWM from runner beans.[79] Using a shorter treatment with chlorite/HOAc, less modified wall glycoproteins have been isolated and some of their structural features elucidated; for details of the method see ref. 12. The HP-poor proteins have generally been found to be more soluble in 1 M KOH[79] and these polymers are proving to be mainly polysaccharide–protein–polyphenol complexes.

It is important to bear in mind that because of the nature of the chemical methods used, some of the structural features of the polysaccharides are bound to be destroyed or modified. For example, the pectins would be partially degraded by the hot aqueous extraction conditions used[150] and the acetyl groups in the xylans of lignified tissues would be saponified during alkaline treatment. However, a proportion of the acetylated xylans could be extracted with DMSO.[135]

5.2.2. CWM FROM TISSUES CONTAINING LITTLE OR NO LIGNIN

If the tissue is (relatively) free of lignin, then we are of the opinion that the delignification stage can be omitted from the scheme shown in Fig. 10. As chlorite/HOAc treatment modifies some of the protein amino acids, deletion of this step has facilitated the detection, isolation and characterisation of polysaccharide–protein–polyphenol complexes from the depectinated CWM of the parenchymatous tissue of potato, runner beans, cabbage and apples (Selvendran, O'Neill and Stevens, unpublished results). Furthermore, in the case of CWM of parenchymatous tissues of mature runner bean pods, the insoluble residue after alkali extraction (α-cellulose) was shown to be composed mainly of cellulose with small but significant amounts of pectic material and HP-rich glycoproteins.[79,146] The HP-rich glycoprotein component could be released from the α-cellulose residue by treatment with chlorite/HOAc.[79] Treatment of α-cellulose with cellulase from *Trichoderma viride* solubilised a high molecular weight component rich in uronic acid; this component accounted for approximately 12% (w/w) of the α-cellulose residue of runner bean.[146] Similar observations on the association of some pectic material with the α-cellulose fraction have been made with cabbage[115] and potato cell walls.[116]

Thus it would appear that some useful information could be obtained on the nature, mode of occurrence and association of polysaccharides, proteins and polyphenol (lignin-like) complexes in the CWM by slightly altering the standard sequential extraction procedures.

5.3. Fractionation of Cell Wall Polymers

5.3.1. Pectic Substances

The pectic substances can be readily separated into neutral and acidic components by either chromatography on DEAE–cellulose[59,151,152] or DEAE–Sephadex columns,[147,153,154] or by electrophoresis on glass fibre paper.[41] Because DEAE–Sephadex columns tend to shrink appreciably on usage, we now prefer to use DEAE–Sephacel columns (Stevens and Selvendran, unpublished results). DEAE–Sephacel is a cellulose ion-exchanger in which the cellulose is regenerated to give a bead-form gel-like structure, and this material shrinks only slightly on usage. The column is eluted at first with buffer, followed by a linear gradient of increasing ionic strength of sodium chloride to 1 M, and the fractions are monitored by reaction with phenol-sulphuric acid. It should be noted that the recoveries of acidic polysaccharides from the anion-exchange columns are usually low, about 60–70% (Stevens and Selvendran, unpublished results).

5.3.2. Alkali-Soluble Polymers

Partial fractionation of the hemicelluloses can be achieved by solvent extraction techniques, using aqueous ethanol mixtures.[155] Such methods would give quantitative recoveries, but the degree of fractionation is usually not adequate. The hemicelluloses may also be fractionated as their acetates,[156] or by chromatography on anion-exchange columns. As with pectic substances, the latter method separates them into neutral and acidic fractions. Chromatography can be effected on either DEAE–cellulose, DEAE–Sephadex or DEAE–Sephacel columns.[157–166] The elution of the polymers is followed by monitoring the reaction with phenol-sulphuric acid, and by measuring the absorption at 280 nm. The measurement of the absorption at 280 nm helps one to screen the fractions for the presence of phenolic material and to detect proteins rich in aromatic amino acids. The fractions from a peak are pooled, dialysed against water and freeze-dried. Further fractionation of the complexes rich in glycoproteins can be effected on a hydroxylapatite column (O'Neill and Selvendran, unpublished results). The 4 M KOH-soluble polymers contain mainly xyloglucans and some acidic components. To purify the xyloglucans it is necessary to remove the acidic components and this can be achieved by fractionation on a DEAE–Sephadex column (formate form).[116] The neutral polysaccharides from the column can be further fractionated by either (a) copper complex formation, as described by Aspinall *et al.*,[167] or (b) by chromatography on a cellulose column using alkali as the eluant. Ring and Selvendran have isolated a xyloglucan from potatoes by the first method and have partially characterised it.[116] O'Neill and Selvendran have attempted further purification of the xyloglucan obtained by the second method, as a borate complex, on a DEAE–Sephacel column.[146]

5.3.3. Hydroxyproline-Rich Glycoproteins

The HP-rich wall glycoproteins (from the chlorite/HOAc extract) can be fractionated into neutral, weakly acidic and strongly acidic components by chromatography on DEAE–Sephadex column. For details of the procedure see ref. 12.

5.4. Characterisation of Cell Wall Polysaccharides

A partial characterisation of a polysaccharide preparation is provided by the method of isolation which may occasionally be so selective as to yield only one single polysaccharide species. Particularly the methods based on differences in charge density, i.e. fractionation on ion-exchange materials, are in themselves a means of characterising the isolated materials. The extent to which it is necessary to characterise a polysaccharide preparation naturally varies with the needs of the problem under study. Qualitative and quantitative analyses of the component sugars and other constituents are often enough to establish the nature of the type of polysaccharide, at least tentatively. Further analysis, including methylation analysis of the undegraded and partially degraded polysaccharide, and the identification of the products obtained on partial acid hydrolysis, enzymatic degradation and acetolysis of the parent polysaccharide, may be carried out only if a more complete picture of the structure of the polysaccharide is required. For an excellent account of the various methods available for the elucidation of the structure of polysaccharides see Aspinall.[35]

5.4.1. Quantitative Analysis of Components

(*a*) *Estimation of Neutral Sugars and Uronic Acids*

The neutral sugars can be released quantitatively from most polysaccharides by Saeman hydrolysis and by hydrolysis with 1 M H_2SO_4 for 2·5 h, converted to alditol acetates and estimated by GLC. In certain cases (e.g. pectins, 4-*O*-methylglucuronoxylans and glucuronoarabinoxylans), the complete hydrolysis of glycosidic linkages, e.g. of glycosiduronic acid linkages, requires very severe conditions and a compromise must be made between maximum hydrolysis and minimum decomposition. The lower rates of hydrolysis of glycosiduronic acid linkage in aldobiouronic acids are attributed to steric factors; the nature of the linkage markedly affects the magnitude of the stabilising effect due to the carboxyl group on C-5.[168] It is useful to note that the D-glucosiduronic linkage in 2-*O*-(4-*O*-methyl-α-D-glucopyranosyluronic acid)-D-xylose is hydrolysed at 1/60 the rate for D-glucosidic linkage in 2-*O*-(4-*O*-methyl-α-D-glucopyranosyl)-D-xylose and at 1/250 to 1/300 of that of xylobiose.[169]

Quantitative estimation of uronic acids present in acidic cell wall polysaccharides is difficult. The galacturonic acid content of pectins can be estimated reasonably accurately by the colorimetric methods discussed earlier. However, colorimetric methods do not give reliable estimates when the uronic acid content of the polysaccharide is low, e.g. the uronic acid content of hemicelluloses. Blake and Richards[54] have concluded that base titrimetric uronic acid estimations give better estimates than the more complex methods involving derivatisation procedures, because the latter type of methodology gives variable results. The reader is advised to look up their paper for a critical assessment of the GLC methods proposed for uronic acid determination. However, GLC methods are essential for the unambiguous differentiation between the different uronic acids and occupy an important position in current methodology. Buchala and Wilkie[170] have proposed a different method. The uronosyl residues in acidic hemicelluloses are first esterified. The uronate residues are then reduced with sodium borotritide and the labelled products determined. It should be noted that in this method, before esterification, the latent aldehyde groups of the end residues in the various hemicelluloses should be reduced with sodium borohydride. As we mentioned earlier, a promising method for water-soluble polyuronides is to reduce the carbodiimide-activated carboxyl groups with sodium borohydride to convert the uronic acid residues in the polysaccharide to the corresponding neutral sugars, thus replacing the acid resistant glycosiduronic acid linkage with the more acid-labile glycosyl bonds.[55] The resulting neutral polysaccharides can then be hydrolysed quantitatively and the sugars analysed by GLC. For an application of this method see ref. 56.

The enantiomeric nature of the sugar residues may be obtained by (1) using pure and specific enzymes now available, and (2) by capillary gas–liquid chromatography of per(trimethylsilyl)ated (−)-2-butyl glycosides of neutral monosaccharides[171] or by GLC of their acetylated glycosides formed from chiral alcohols.[172]

(b) Methylation Analysis and Periodate Oxidation Studies of Polysaccharides

Methylation analysis is an important tool used for the elucidation of the structure of polysaccharides. It involves replacing all free hydroxyl groups in the polysaccharide by —OCH_3 groups, hydrolysis of the fully methylated polysaccharide to a mixture of partially methylated sugars, and qualitative and quantitative analysis of this mixture. The free hydroxyls in the partially methylated sugars indicate the positions in which the sugar residues are substituted. The Hakomori methylation procedure[173,174] and the use of mass spectrometers to examine the sugar derivatives have considerably increased the scope of structural studies on polysaccharides. Lindberg and his collaborators[175,176] have converted the partially methylated sugars into partially methylated alditol acetates and identified these derivatives by gas–liquid chromatography–mass spectrometry (GC–MS). Several groups of workers have used this technique to partially characterise the cell wall polysaccharides from a range of tissues.[25,52,116,127,141–145,167,177,178] The method is shown in Fig. 11. In this scheme it should be noted that an estimate of the galacturonosyl residues (of pectins) was obtained from an analysis of the methylated fraction after reduction with $LiAlH_4$ or $LiAlD_4$. This is because the galacturonosyl residues do not form a sufficiently volatile derivative before reduction. The method provides information on the mode of

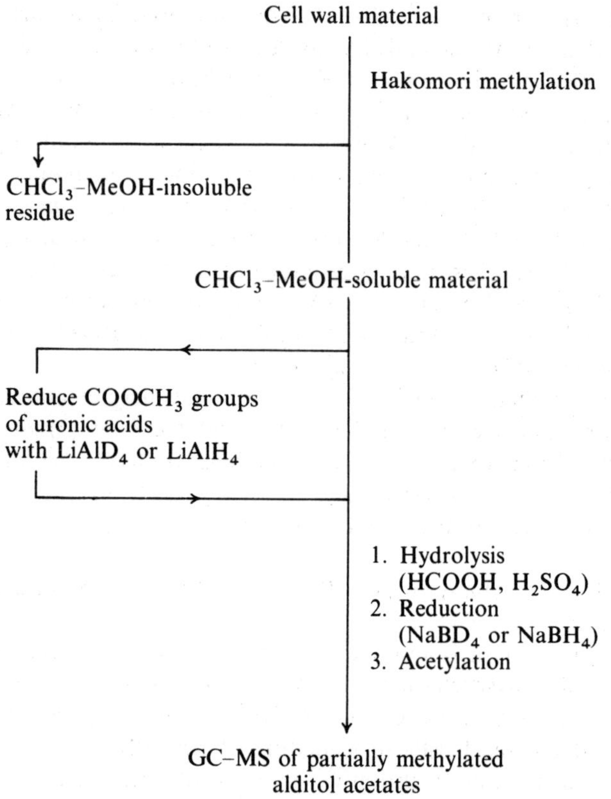

FIG. 11. The preparation, for GC–MS, of partially methylated alditol acetates of sugars and uronic acids of cell wall polysaccharides.

occurrence of the various sugar residues in the polysaccharide, but not on their mutual arrangement or on the anomeric nature of the sugar residues.

The mode of occurrence of sugar residues which are very susceptible to acid hydrolysis (e.g. furanosidic sugars or terminal fucose residues) can be obtained from methylation analysis of the undegraded and partially degraded polysaccharides. Likewise, the methylated polysaccharides may be partially hydrolysed and re-methylated using trideuteriomethyl iodide. This method yields similar but more detailed information on the positions of the acid-labile sugars. In our laboratory, this technique has been used to determine the site of attachment of arabinofuranoside residues in wheat bran arabinoxylans[52] and in potato xyloglucan.[116]

Confirmatory evidence on the structure of polysaccharides can be obtained by periodate oxidation studies. The reaction of polysaccharides with periodate may be followed by determining the amount of (a) periodate consumed, (b) formic acid generated and (c) surviving sugar residues.[179,180] The periodate-oxidised polysaccharides may be examined further, preferably after reduction of the polyaldehyde to the polyalcohol, followed by hydrolysis.[180,181] Controlled hydrolysis of the polyalcohol results in the preferential hydrolysis of acyclic acetal linkages with the liberation of glycosides of polyhydric alcohols.[180]

(c) Identification of Oligosaccharides from Polysaccharides

Information about the mutual arrangement of sugar residues in a polysaccharide can be obtained by the following methods: (1) By characterising the oligosaccharides released on partial acid hydrolysis,[52] enzymatic hydrolysis[116] or acetolysis[146,167] of the polysaccharide. It is now possible to sequence the sugar residues in an oligosaccharide, after suitable derivatisation procedures, by GC–MS. The method has been used by our group for sequencing oligosaccharides containing up to four sugar residues.[52,116,146] Derivatives of higher oligosaccharides can be sequenced by using direct insertion mass spectrometry.[182] (2) By characterising the partially methylated oligosaccharides produced on partial acid hydrolysis of the methylated polysaccharide. This method has been used by Valent *et al.*[183] for sequencing complex carbohydrates. An account of the applications of GC–MS for cell wall analysis is given in ref. 184.

6. CONCLUDING REMARKS

Since many people in various laboratories are concerned with developing new and improved methods for analysing dietary fibre, much pertinent information will become available in the future and will augment the present discussion. However, an effort has been made to organise the relevant information on the structure and properties of dietary fibre from fresh and processed foods. It is hoped that this information will be intelligible and useful to the laboratory worker who has some experience with the analysis of polysaccharides.

The goal of this chapter will have been realised insofar as these researchers are helped to isolate and fractionate DF into its constituent polymers in order to elucidate further the role of DF in human nutrition and health. This is important because the mechanisms whereby cereal fibre and vegetable/fruit fibre affect the human colon are different.[4,185] The structures and solubility characteristics of the DF polymers from wheat bran[52,148,186] are different from those of cabbage,[115,187] apples, potatoes[116] and runner beans.[79,146] Whereas the former contains mainly arabinoxylans, cellulose, some β-D-glucan and lignin,[138,157,161,178] the latter contain mainly pectic substances, cellulose and some hemicellulosic polysaccharides, especially xyloglucans. Phenolic ester cross-linkages are prominent in the bran fibre but not in the fibre from the other materials. These differences appear to correlate with the greater influence of cereal fibre, as opposed to vegetable fibre, on faecal bulking. Some of these aspects are discussed at length in ref. 187. Furthermore, the detailed structural studies give useful pointers for developing a hierarchy of simpler, though less informative, procedures for use in epidemiological studies and in routine food analysis for quality control and labelling purposes.

REFERENCES

1. TROWELL, H. (1972). *Am. J. Clin. Nutr.*, **25**, 926.
2. TROWELL, H., SOUTHGATE, D. A. T., WOLEVER, T. M. S., LEEDS, A. R., GASSUL, M. A. and JENKINS, D. A. (1976). *Lancet*, **1**, 967.
3. SALYERS, A. A., VERCELLOTTI, J. R., WEST, S. E. H. and WILKINS, T. D. (1977). *Appl. Environ. Microbiol.*, **33**, 319.
4. STEPHEN, A. M. and CUMMINGS, J. H. (1980). *Nature (London)*, **284**, 283.
5. NORTHCOTE, D. H. (1963). *Symp. Soc. Exp. Biol.*, **17**, 157.
6. NORTHCOTE, D. H. (1972). *Annu. Rev. Plant Physiol.*, **23**, 113.
7. ASPINALL, G. O. (1980). In *The Biochemistry of Plants*, Vol. 3: *Carbohydrates: Structure and Function*, Preiss, J. (Ed.), Academic Press, New York and London, p. 473.
8. SOUTHGATE, D. A. T. (1976). In *Fiber in Human Nutrition*, Spiller, G. A. and Amen, R. J. (Eds), Plenum Press, New York and London, p. 31.
9. SELVENDRAN, R. R. (1983). In *Dietary Fibre*, Birch, G. G. and Parker, K. J. (Eds), Applied Science Publishers, London, p. 95.
10. LAMPORT, D. T. A. (1970), *Annu. Rev. Plant Physiol.*, **21**, 235.
11. SELVENDRAN, R. R. (1975). *Phytochemistry*, **14**, 1011.
12. O'NEILL, M. A. and SELVENDRAN, R. R. (1980). *Biochem. J.*, **187**, 53.
13. SARKANEN, K. V. and HERGERT, H. L. (1971). In *Lignins: Occurrence, Formulation, Structure and Reactions*, Sarkanen, K. V. and Ludwig, C. H. (Eds), Wiley-Interscience, New York and London, p. 43.
14. LAI, Y. Z. and SARKANEN, K. V. (1971). In *Lignins: Occurrence, Formulation, Structure and Reactions*, Sarkanen, K. V. and Ludwig, C. H. (Eds), Wiley-Interscience, New York and London, p. 165.
15. SELVENDRAN, R. R. and DUPONT, M. S. (1980). *Cereal Chem.*, **57**, 278.

16. SELVENDRAN, R. R. and DUPONT, M. S. (1980). *J. Sci. Fd Agric.*, **31**, 1173.
17. SOUTHGATE, D. A. T., HUDSON, G. J. and ENGLYST, H. (1978). *J. Sci. Fd Agric.*, **29**, 979.
18. THEANDER, O. and AMAN, P. (1979). *Swedish J. Agric. Res.*, **9**, 97.
19. SCHWEIZER, T. F. and WÜRSCH, P. (1979). *J. Sci. Fd Agric.*, **30**, 613.
20. VAN SOEST, P. J. and WINE, R. H. (1967). *J. Ass. Off. Anal. Chem.*, **50**, 50.
21. ROBERTSON, J. B. and VAN SOEST, P. J. (1977). *J. Animal Sci.*, **45**, Suppl. 1, 254.
22. ROBERTSON, J. B. and VAN SOEST, P. J. (1981). In *The Analysis of Dietary Fiber in Food*, James, W. P. T. and Theander, O. (Eds), Marcel Dekker, New York and Basel, p. 123.
23. SELVENDRAN, R. R., RING, S. G. and DUPONT, M. S. (1981). In *The Analysis of Dietary Fiber in Food*, James, W. P. T. and Theander, O. (Eds), Marcel Dekker, New York and Basel, p. 95.
24. ENGLYST, H., WIGGINS, H. S. and CUMMINGS, J. H. (1982). *The Analyst*, **107**, 307.
25. RING, S. G. and SELVENDRAN, R. R. (1978). *Phytochemistry*, **17**, 745.
26. ENGLYST, H. (1981). In *The Analysis of Dietary Fiber in Food*, James, W. P. T. and Theander, O. (Eds), Marcel Dekker, New York and Basel, p. 71.
27. TERRY, R. A. and OUTEN, G. E. (1973). *Chem. & Ind.*, p. 1116.
28. SCHALLER, D. (1973). *Am. J. Clin. Nutr.*, **31**, S99.
29. MCQUEEN, R. E. and NICHOLSON, J. W. G. (1979). *J. Ass. Off. Anal. Chem.*, **62**, 676.
30. MARTLETT, J. A. and LEE, S. C. (1980). *J. Fd Sci.*, **45**, 1688.
31. LEE, S., KIVILAAN, A. and BANDURSKI, R. S. (1967). *Plant Physiol.*, **42**, 968.
32. KIVILAAN, A., BANDURSKI, R. S. and SCHULZE, A. (1971). *Plant Physiol.*, **48**, 389.
33. SELVENDRAN, R. R., MARCH, J. F. and RING, S. G. (1979). *Anal. Biochem.*, **96**, 282.
34. ADAMS, G. A. (1965). In *Methods in Carbohydrate Chemistry*, Vol. 5, Whistler, R. L. and BeMiller, J. N. (Eds), Academic Press, New York and London, p. 269.
35. ASPINALL, G. O. (1973). In *Techniques of Chemistry*, Vol. 4: *Elucidation of Organic Structures by Physical and Chemical Methods*, Bentley, K. W. and Kirby, G. W. (Eds), Wiley-Interscience, New York and London, p. 379.
36. BAILEY, R. W. (1973). In *Chemistry and Biochemistry of Herbage*, Vol. 1, Butler, G. W. and Bailey, R. W. (Eds), Academic Press, London and New York, p. 157.
37. DUTTON, G. G. S. (1973). In *Advances in Carbohydrate Chemistry and Biochemistry*, Vol. 28, Tipson, R. S. and Horton, D. (Eds), Academic Press, New York and London, p. 11.
38. CONRAD, H. E., BAMBURG, J. R., EPLEY, J. D. and KINDT, T. J. (1966). *Biochemistry*, **5**, 2808.
39. FEATHER, M. S. and HARRIS, J. F. (1965). *J. Org. Chem.*, **30**, 153.
40. JERMYN, M. A. and ISHERWOOD, F. A. (1956). *Biochem. J.*, **64**, 123.
41. BARRETT, A. J. and NORTHCOTE, D. H. (1965). *Biochem. J.*, **94**, 617.
42. NEUBERGER, A. and MARSHALL, R. D. (1966). In *Glycoproteins*, Gottschalk, A. (Ed.), Elsevier, Amsterdam, p. 190.

43. ALBERSHEIM, P., NEVINS, D. J., ENGLISH, P. D. and KARR, A. (1967). *Carbohyd. Res.*, **5**, 340.
44. HOUGH, L., JONES, J. V. S. and WUSTEMAN, P. (1972). *Carbohyd. Res.*, **21**, 9.
45. KIM, J. H., SHOME, B., LIA, O. T. H. and PIERCE, J. C. (1967). *Anal. Biochem.*, **20**, 258.
46. PAINTER, T. J. (1960). *Chem. & Ind.*, p. 1214.
47. HOUGH, L. and PRIDHAM, J. B. (1959). *Biochem. J.*, **73**, 550.
48. CHAMBERS, R. E. and CLAMP, J. R. (1971). *Biochem. J.*, **125**, 1009.
49. RUBERY, P. H. and NORTHCOTE, D. H. (1970). *Biochim. Biophys. Acta*, **222**, 95.
50. WHISTLER. R. L. and RICHARDS, G. N. (1958). *J. Am. Chem. Soc.*, **80**, 4888.
51. ROY, N. and TIMELL, T. E. (1968). *Carbohyd. Res.*, **6**, 488.
52. RING, S. G. and SELVENDRAN, R. R. (1980). *Phytochemistry*, **19**, 1723.
53. BLAKE, J. D. and RICHARDS, G. N. (1968). *Carbohyd. Res.*, **8**, 275.
54. BLAKE, J. D. and RICHARDS, G. N. (1970). *Carbohyd. Res.*, **14**, 375.
55. TAYLOR, R. L. and CONRAD, H. E. (1972). *Biochemistry*, **11**, 1383.
56. ASPINALL, G. O. and JIANG, K. (1974). *Carbohyd. Res.*, **38**, 247.
57. SMITH, F. and MONTGOMERY, R. (1956). In *Methods of Biochemical Analysis*, Vol. 3, Glick, D. (Ed.), Interscience, New York and London, p. 153.
58. HODGE, J. E. and HOFREITER, B. T. (1962). In *Methods in Carbohydrate Chemistry*, Vol. 1, Whistler, R. L. and Wolfrom, M. L. (Eds), Academic Press, New York and London, p. 380.
59. PARK, J. T. and JOHNSON, N. J. (1949). *J. Biol. Chem.*, **181**, 149.
60. HEWITT, L. F. (1938). *Biochem. J.*, **32**, 1554.
61. DISCHE, Z. (1962). In *Methods in Carbohydrate Chemistry*, Vol. 1, Whistler, R. L. and Wolfrom, M. L. (Eds), Academic Press, New York and London, p. 477.
62. ASHWELL, G. (1957). In *Methods in Enzymology*, Vol. 3, Colowick, S. P. and Kaplan, N. A. (Eds), Academic Press, New York, p. 73.
63. SPIRO, R. G. (1966). In *Methods in Enzymology*, Vol. 8: *Complex Carbohydrates*, Neufeld, E. F. and Ginsburg, V. (Eds), Academic Press, New York and London, p. 3.
64. SELVENDRAN, R. R., RING, S. G. and DUPONT, M. S. (1979). *Chem. & Ind.*, p. 225.
65. AVIGAD, G. (1975). In *Methods in Enzymology*, Vol. 41, *Carbohydrate Metabolism*, Part B, Wood, W. A. (Ed.), Academic Press, New York and London, p. 31.
66. HOUGH, L. and JONES, J. K. N. (1962). In *Methods in Carbohydrate Chemistry*, Vol. 1, Whistler, R. L. and Wolfrom, M. L. (Eds), Academic Press, New York and London, p. 21.
67. JERMYN, M. A. and ISHERWOOD, F. A. (1949). *Biochem. J.*, **44**, 402.
68. DUBOIS, M., GILLES, K. A., HAMILTON, J. K., REBERS, P. A. and SMITH, F. (1956). *Anal. Chem.*, **28**, 350.
69. FRANÇOIS, C., MARSHALL, R. D. and NEUBERGER, A. (1962). *Biochem. J.*, **83**, 335.
70. JERMYN, M. A. (1955). In *Modern Methods of Plant Analysis*, Vol. 2, Paech, K. and Tracey, M. V. (Eds), Springer-Verlag, Berlin, p. 197.
71. THORNBER, J. P. and NORTHCOTE, D. H. (1961). *Biochem. J.*, **81**, 455.

72. SELVENDRAN, R. R., PERERA, B. P. M. and SELVENDRAN, S. (1972). *J. Sci. Fd Agric.*, **23**, 1119.
73. OHMS, J. I., ZEC. J., BENSON, J. V. and PATTERSON, J. A. (1967). *Anal. Biochem.*, **20**, 51.
74. LEE, Y. C., MCKELVY, J. F. and LANG, D. (1969). *Anal. Biochem.*, **27**, 567.
75. WALBORG, E. F. (1969). *Anal. Biochem.*, **29**, 433.
76. SPIRO, R. G. (1972). In *Methods in Enzymology*, Vol. 28: *Complex Carbohydrates*, Part B, Ginsburg, V. (Ed.), Academic Press, New York and London, p. 3.
77. DAVIES, A. M. C., ROBINSON, D. S. and COUCHMAN, R. (1974). *J. Chromatogr.*, **101**, 307.
78. SELVENDRAN, R. R., DAVIES, A. M. C. and TIDDER, E. (1975). *Phytochemistry*, **14**, 2169.
79. SELVENDRAN, R. R. (1975). *Phytochemistry*, **14**, 2175.
80. BISHOP, C. T. (1962). In *Methods of Biochemical Analysis*, Vol. 10, Glick, D. (Ed.), Interscience, New York and London, p. 1.
81. SWEELEY, C. C., BENTLEY, R., MALUTA, M. and WELLS, W. W. (1963). *J. Am. Chem. Soc.*, **85**, 2497.
82. CLAMP, J. R. (1967). *Biochim. Biophys. Acta*, **147**, 342.
83. ABDEL-AKHER, M., HAMILTON, J. K. and SMITH, F. (1951). *J. Am. Chem. Soc.*, **73**, 4691.
84. BISHOP, C. T. (1960). *Can. J. Chem.*, **38**, 1636.
85. SAWARDEKER, J. S., SLONEKER, J. H. and JEANES, A. (1965). *Anal. Chem.*, **37**, 1602.
86. CROWELL, E. P. and BURNETT, B. B. (1967). *Anal. Chem.*, **39**, 121.
87. SHAPIRA, J. (1969). *Nature (London)*, **222**, 792.
88. VARMA, R., VARMA, R. S. and WARDI, A. H. (1973). *J. Chromatogr.*, **77**, 222.
89. MORRISON, I. M. (1975). *J. Chromatogr.*, **108**, 361.
90. JONES, T. M. and ALBERSHEIM, P. (1972). *Plant Physiol.*, **49**, 926.
91. BUCHALA, A. J., FRASER, C. G. and WILKIE, K. C. B. (1971). *Phytochemistry*, **10**, 1285.
92. GORDON, A. H., HAY, A. J., DINSDALE, D. and BACON, J. S. D. (1977). *Carbohyd. Res.*, **57**, 235.
93. HYAKUTAKE, H. and HANAI, T. (1975). *J. Chromatogr.*, **108**, 385.
94. LINDEN, J. C. and LAWHEAD, C. L. (1975). *J. Chromatogr.*, **105**, 125.
95. CONRAD, E. C. and PALMER, J. K. (1976). *Fd Technol.*, **30** (Oct.), 84.
96. LADISCH, M. R. and TSAO, G. T. (1978). *J. Chromatogr.*, **166**, 85.
97. VERHAAR, L. A. and KUSTER, B. F. M. (1981). *J. Chromatogr.*, **210**, 279.
98. VORAGEN, A. G. T., SCHOLS, H. A., DE VRIES, J. A. and PILNIK, W. (1982). *J. Chromatogr.*, **244**, 327.
99. BROWNING, B. L. (1967). *Methods of Wood Chemistry*, Vol. 11, Interscience, New York and London, p. 785.
100. PEARL, I. A. (1967). *The Chemistry of Lignin*, Edward Arnold, London; Marcel Dekker, New York, p. 37.
101. VAN SOEST, P. J. and WINE, R. H. (1968). *J. Am. Off. Anal. Chem.*, **51**, 780.
102. JOHNSON, D. B., MOORE, W. E. and ZANK, C. (1961). *Tappi*, **44**, 793.
103. MORRISON, I. M. (1972). *J. Sci. Fd Agric.*, **23**, 455.
104. MORRISON, I. M. (1972). *J. Sci. Fd Agric.*, **23**, 1463.

105. WHITEHEAD, D. L. and QUICKE, G. V. (1964). *J. Sci. Fd Agric.*, **15**, 417.
106. CHANG, H.-M. and ALLAN, G. G. (1971). In *Lignins: Occurrence, Formulation, Structure and Reactions*, Sarkanen, K. V. and Ludwig, C. H. (Eds), Wiley-Interscience, New York and London, p. 433.
107. VAN SOEST, P. J. (1963). *J. Am. Off. Anal. Chem.*, **46**, 829.
108. SELVENDRAN, R. R. (1978). *Chem. & Ind.*, p. 428.
109. HARTLEY, R. D. (1973). *Phytochemistry*, **12**, 661.
110. COLLINGE, S. K., GROSH, R. S. and MAHONEY, A. W. (1980). *J. Fd Biochem.*, **4**, 111.
111. SCHALLER, D. R. (1981). *Cereal Fds Wld*, **26**, 295.
112. SOUTHGATE, D. A. T. (1969), *J. Sci. Fd Agric.*, **20**, 331.
113. SOUTHGATE, D. A. T. (1976). In *Fiber in Human Nutrition*, Spiller, G. A. and Amen, R. J. (Eds), Plenum Press, New York and London, p. 73.
114. SOUTHGATE, D. A. T. (1981). In *The Analysis of Dietary Fiber in Food*, James, W. P. T. and Theander, O. (Eds), Marcel Dekker, New York and Basel, p. 1.
115. STEVENS, B. J. H. and SELVENDRAN, R. R. (1980). *J. Sci. Fd Agric.*, **31**, 1257.
116. RING, S. G. and SELVENDRAN, R. R. (1981). *Phytochemistry*, **20**, 2511.
117. LAINE, R. A., VARO, P. and KOIVISTOINEN, P. E. (1981). In *The Analysis of Dietary Fiber in Food*, James, W. P. T. and Theander, O. (Eds), Marcel Dekker, New York and Basel, p. 21.
118. THEANDER, O. and AMAN, P. (1981). In *The Analysis of Dietary Fiber in Food*, James, W. P. T. and Theander, O. (Eds), Marcel Dekker, New York and Basel, p. 51.
119. SCHWEIZER, T. F. and WÜRSCH, P. (1981). In *The Analysis of Dietary Fiber in Food*, James, W. P. T. and Theander, O. (Eds), Marcel Dekker, New York and Basel, p. 203.
120. GREENWOOD, C. T. and MILNE, E. A. (1968). In *Advances in Carbohydrate Chemistry*, Vol. 23, Wolfrom, M. L. and Tipson, R. S. (Eds), Academic Press, New York and London, p. 281.
121. ASPINALL, G. O., COTTRELL, I. W., MOLLOY, J. A. and UDDIN, M. (1970). *Can. J. Chem.*, **48**, 1290.
122. BLUMENKRANTZ, N. and ASBOE-HANSEN, G. (1973). *Anal. Biochem.*, **54**, 484.
123. KINTNER, P. K. and VAN BUREN, J. P. (1982) *J. Fd. Sci.*, **47**, 756.
124. DISCHE, Z. (1962). In *Methods in Carbohydrate Chemistry*, Vol. 1, Whistler, R. L. and Wolfrom, M. L. (Eds), Academic Press, New York and London, p. 487.
125. DISCHE, Z. (1962). In *Methods in Carbohydrate Chemistry*, Vol. 1, Whistler, R. L. and Wolfrom, M. L. (Eds), Academic Press, New York and London, p. 490.
126. MARSHALL, J. J. (1974). In *Advances in Carbohydrate Chemistry and Biochemistry*, Vol. 30, Tipson, R. S. and Horton, D. (Eds), Academic Press, New York and London, p. 257.
127. O'NEILL, M. A. and SELVENDRAN, R. R. (1980). *Carbohyd. Res.*, **79**, 115.
128. CHEN, W. J. L. and ANDERSON, J. W. (1981). *Am. J. Clin. Nutr.*, **34**, 1077.
129. HELLENDOORN, E. W., NOORDHOFF, M. G. and SLAGMAN, J. (1975). *J. Sci. Fd Agric.*, **26**, 1461.
130. HELLENDOORN, E. W. (1981). *Am. J. Clin. Nutr.*, **34**, 1437.
131. HONIG, D. H. and RACKIS, J. J. (1979). *J. Agric. Fd Chem.*, **27**, 1262.

132. (a) Asp, N. G. and Johansson, C. G. (1981). In *The Analysis of Dietary Fiber in Food*, James, W. P. T. and Theander, O. (Eds), Marcel Dekker, New York and Basel, p. 173; (b) Asp, N. G., Johansson, C. G., Hallmer, H. and Siljestrom, M. (1983). *J. Agric. Fd Chem.*, **31**, 476.
133. Jansson, P. E., Kenne, L. and Lindberg, B. (1975). *Carbohyd. Res.*, **45**, 275.
134. Klose, R. E. and Glicksman, M. (1972). In: *Handbook of Food Additives*, 2nd Edn, Furia, T. E. (Ed.), CRC Press, Ohio, Chap. 7, pp. 295–359.
135. Hagglund, E., Lindberg, B. and McPherson, J. (1956). *Acta Chem. Scand.*, **10**, 1160.
136. Morrison, I. M. (1973). *Phytochemistry*, **12**, 2979.
137. Hoff, J. E. and Castro, M. D. (1969). *J. Agric. Fd Chem.*, **17**, 1328.
138. Mares, D. J. and Stone, B. A. (1973). *Aust. J. Biol. Sci.*, **26**, 793.
139. Ballance, G. M. and Manners, D. J. (1978). *Carbohyd. Res.*, **61**, 107.
140. Anderson, R. L. and Stone, B. A. (1973). *Aust. J. Biol. Sci.*, **26**, 135.
141. Talmadge, K. W., Keegstra, K., Bauer, W. D. and Albersheim, P. (1973). *Plant Physiol.*, **51**, 158.
142. Bauer, W. D., Talmadge, K. W., Keegstra, K. and Albersheim, P. (1973). *Plant Physiol.*, **51**, 174.
143. Keegstra, K., Talmadge, K. W., Bauer, W. D. and Albersheim, P. (1973). *Plant Physiol.*, **51**, 188.
144. Wilder, B. M. and Albersheim, P. (1973). *Plant Physiol.*, **51**, 889.
145. Albersheim, P., Bauer, W. D., Keegstra, K. and Talmadge, K. W. (1973). In *Biogenesis of Plant Cell Wall Polysaccharides*, Loewus, F. (Ed.), Academic Press, New York, p. 117.
146. O'Neill, M. A. and Selvendran, R. R. (1983). *Carbohyd. Res.*, **111**, 239.
147. Stevens, B. J. H. and Selvendran, R. R. (1980). *Phytochemistry*, **19**, 559.
148. Selvendran, R. R., Ring, S. G., O'Neill, M. A. and DuPont, M. S. (1980). *Chem. & Ind.*, p. 885.
149. Blake, J. D., Murphy, P. T. and Richards, G. N. (1971). *Carbohyd. Res.*, **16**, 49.
150. Aspinall, G. O. (1970). *Polysaccharides*, Pergamon Press, Oxford, p. 116.
151. Knee, M. (1970). *J. Exp. Bot*, **21**, 651.
152. Siddiqui, I. R. and Wood, P. J. (1976). *Carbohyd. Res.*, **50**, 97.
153. Konno, H., Yamasaki, Y. and Ozawa, J. (1980). *J. Agric. Biol. Chem.*, **44**, 2195.
154. Aspinall, G. O., Molloy, J. A. and Craig, J. W. T. (1969). *Can. J. Biochem.*, **47**, 1063.
155. Reid, J. S. G. and Wilkie, K. C. B. (1969). *Phytochemistry*, **8**, 2045.
156. Montgomery, R. and Smith, F. (1955). *J. Am. Chem. Soc.*, **77**, 2834.
157. Mares, D. J. and Stone, B. A. (1973). *Aust. J. Biol. Sci.*, **26**, 813.
158. Reid, J. S. G. and Wilkie, K. C. B. (1969). *Phytochemistry*, **8**, 2045.
159. Reid, J. S. G. and Wilkie, K. C. B. (1969). *Phytochemistry*, **8**, 2053.
160. Reid, J. S. G. and Wilkie, K. C. B. (1969). *Phytochemistry*, **8**, 2059.
161. Medcalf, B., d'Appolonia, B. L. and Gilles, K. A. (1968). *Cereal Chem.*, **45**, 539.
162. d'Appolonia, B. L. and Macarthur, L. A. (1976). *Cereal Chem.*, **53**, 711.
163. Naivikul, O. and d'Appolonia, B. L. (1979). *Cereal Chem.*, **56**, 45.

164. NEUKOM, H. and KUNDIG, W. (1965). In *Methods in Carbohydrate Chemistry*, Vol. 5, Whistler, R. L. (Ed.), Academic Press, New York and London, p. 14.
165. MORRISON, I. M. (1974). *Carbohyd. Res.*, **36**, 45.
166. SIDDIQUI, I. R. and WOOD, P. J. (1972). *Carbohyd. Res.*, **24**, 1.
167. ASPINALL, G. O., KRISHNAMURTHY, T. N. and ROSELL, K. G. (1977). *Carbohyd. Res.*, **55**, 11.
168. JOHANSSON, T., LINDBERG, B. and THEANDER, O. (1963). *Acta Chem. Scand.*, **17**, 2019.
169. ROY, N. and TIMELL, T. E. (1968). *Carbohyd. Res.*, **6**, 488.
170. BUCHALA, A. J. and WILKIE, K. C. B. (1973). *Phytochemistry*, **12**, 655.
171. GERWIG, G. J., KAMERLING, J. P. and VLIGENTHART, J. F. G. (1978). *Carbohyd. Res.*, **62**, 349.
172. LEONTEIN, K., LINDBERG, B. and LÖNNGREN, J. (1978). *Carbohyd. Res.*, **62**, 359.
173. HAKOMORI, S. (1964). *J. Biochem.* (*Tokyo*), **55**, 205.
174. SANDFORD, P. A. and CONRAD, H. E. (1966). *Biochemistry*, **5**, 1508.
175. LINDBERG, B. (1972). In *Methods in Enzymology*, Vol. 28, Ginsburg, V. (Ed.), Academic Press, New York, p. 178.
176. JANSSON, P. E., KENNE, L., LIEDGREEN, H., LINDBERG, B. and LÖNNGREN, J. (1976). *A Practical Guide to the Methylation Analysis of Carbohydrates*, Chem. Commun. Univ. Stockholm, No. 8.
177. GILKES, N. R. and HALL, M. A. (1977). *New Phytol.*, **78**, 1.
178. BACIC, A. and STONE, B. A. (1981). *Aust. J. Plant Physiol.*, **8**, 475.
179. HAY, G. W., LEWIS. B. A. and SMITH, F. (1965). In *Methods in Carbohydrate Chemistry*, Vol. 5, Whistler, R. L. (Ed.), Academic Press, New York and London, p. 357.
180. GOLDSTEIN, I. J., HAY, G. W., LEWIS, B. A. and SMITH, F. (1965). In *Methods in Carbohydrate Chemistry*, Vol. 5, Whistler, R. L. (Ed.), Academic Press, New York and London, p. 361.
181. ABDEL-AKHER, M., HAMILTON, J. K., MONTGOMERY, R. and SMITH, F. (1952). *J. Am. Chem. Soc.*, **74**, 4970.
182. ASHFORD, D., DESAI, N. N., ALLEN, A. K., NEUBERGER, A., O'NEILL, M. A. and SELVENDRAN, R. R. (1982), *Biochem. J.*, **201**, 199.
183. VALENT, B. S., DARVILL, A. G., MCNEIL, M., ROBERTSON, B. K. and ALBERSHEIM, P. (1980). *Carbohyd. Res.*, **79**, 165.
184. SELVENDRAN, R. R. (1983). In *Recent Developments in Mass Spectrometry in Biochemistry, Medicine and Environmental Research—8*, Frigerio, A. (Ed.), Elsevier, Amsterdam, p. 159.
185. CUMMINGS, J. H., BRANCH, W., JENKINS, D. J. A., SOUTHGATE, D. A. T., HOUSTON, H. and JAMES, W. P. T. (1978). *Lancet*, **1**, 5.
186. BRILLOUET, J. M. and MERCIER, C. (1981). *J. Sci. Fd Agric.*, **32**, 243.
187. STEVENS, B. J. H. and SELVENDRAN, R. R. (1982). *Lebensm.-Wiss. Technol.*, **14**, 301.

Chapter 2

TRACE ELEMENT ANALYSIS

WAYNE R. WOLF and JAMES M. HARNLY

United States Department of Agriculture,
Beltsville, Maryland, USA

'TRACE: An amount of a chemical constituent
not quantitatively determined because of minuteness.'
(*Webster's Dictionary*)

1. INTRODUCTION

With increased awareness of the role of many trace elements in health, a demand has been created for more information on the levels of these elements in foods and beverages. For most people, the intake of foods and beverages constitutes the major source of exposure to these elements. Data on the elemental composition of foods are of interest to nutritionists and toxicologists alike. These data are necessary to establish adequate dietary intakes and to reduce exposure to harmful elements.

The ultimate goal is to develop an information base which gives us a complete understanding and accurate knowledge of (1) human nutritional requirements and toxicity levels, (2) nutrient composition of and the flow of elements through the food supply, and (3) available sources of 'safe' foods which provide required nutrients and are free of harmful effects.[1] At the present time, the extent and quality of the inorganic composition data for foods and beverages are far from satisfactory. These insufficiencies exist because of inadequate analytical methodology, ignorance of the importance of many of these elements (until recently), and the large and constantly changing food supply. The demands for these data place pressure on modern analytical methodology for instrumentation capable of more sensitive and more rapid determinations.

Elements of health interest have historically been divided into two major groups based upon concentration: the 'mineral' elements (a misnomer since they are neither consumed nor function as 'minerals', with the exception of bone formation), which occur at levels above approximately 10 $\mu g/g$, and the 'trace' elements which historically existed at the limits of analytical capability ($<10\ \mu g/g$). Modern analytical methodology has pushed the analytical detection limits down to the sub-parts per billion range and has allowed us to consider the concept of discussing 'quantitative trace analysis' which, as seen from the definition of 'trace' (see above), is impossible. This chapter will thus attempt to deal with the impossible and describe some of the more recent advances in quantitative trace analyses.

Because of the higher level of concentration of the 'mineral' elements, their role in human health is better defined. Nutritionally, recommended dietary intakes have been established, and the major food sources identified. Toxicologically, the presence of many elements at these levels can be harmful. Toxic effects of the elements are generally well defined, and often only qualitative analytical identification of their presence is required. There are a number of analytical methods available which can provide adequate determinations with regard to satisfactory accuracy and precision. The analytical problems are usually those of properly defining the required accuracy and precision and establishing proper quality assurance procedures to attain the required data.

Trace elements on the other hand represent much less defined biological concerns and more difficult analytical challenges. In many cases, the nutritionists and toxicologists are concerned with the same trace elements as shown in Table 1. Many of the trace elements exhibit dual biochemical effects,[3] where the difference between deficiency and toxicity may be only one or two orders of magnitude of concentration. Consequently, it is no longer sufficient to identify the presence of these elements qualitatively. Instead it is necessary to establish quantitatively the levels of deficiency and toxicity, and to characterise the 'normal' range of variation. Quantitative

TABLE 1

TRACE INORGANIC ELEMENTS OF INTEREST IN HUMAN HEALTH[2]

Elements of nutritional interest:
As, Co, Cr, Cu, F, Fe, I, Mn, Mo, Ni, Se, Si, Sn, V, Zn
Elements of toxic interest:
As, Be, Cd, Co, Cr, F, Hg, Mn, Mo, Ni, Pb, Pd, Se, Sn, Tl, V, Zn

analyses at the 'trace' level are a continuous challenge to modern analytical methods.

Of the modern analytical methodology available, the two techniques which generally have the required sensitivity and most potential for accurate trace element analysis are atomic spectrometry and neutron activation analysis. Owing to the widespread and growing use of atomic spectrometry and as a reflection of the authors' bias and experience, this chapter will deal predominantly with atomic spectrometric methods for the analysis of foods. Other techniques will be discussed briefly in Section 4.

Most analytical techniques can measure only the total amount of the inorganic element present. In the future, more and more attention will be directed to identifying the different chemical species that exist for the element in foods. Analytical quantitation of each species will become increasingly important as more knowledge is gained regarding the differing biological roles of each species. For many of the elements, the total amount of element is not of importance biochemically whereas the level of some particular species, which may be in small proportion to the whole amount, may be critically important. This points towards the need to develop even more sensitive and more selective analytical methods for speciation of the trace elements.

2. GENERAL REVIEWS

2.1. Methodology

A number of other works have recently reviewed trace element analysis and there is an extensive body of literature available on the determination of elements in foods and beverages. In addition, there is a considerable volume of literature dealing with the determination of elements in biological and environmental samples which can be applied directly, or with slight modification, to foods. Research on the determination of elements in foods and biological systems is progressing rapidly.[2] New developments are being reported continually. This section is aimed at providing the reader with direction to general reviews (when possible) and to those sources which will allow the reader to become aware of the latest developments.

Several general reviews have appeared on the analysis of trace elements in foods[4,5] and in biological materials.[6–12] Three excellent reviews have been published dealing with the application of specific atomic spectrometric methods to the analysis of foods. Fricke *et al.*[13] and Ihnat[14] reviewed the use of flame atomic absorption spectrometry while Jones and Boyer[15] considered the application of inductively coupled plasma atomic emission spectrometry to the determination of elements in foods.

The present authors have recently discussed in some detail many aspects of atomic spectroscopy as related to determination of inorganic elements in foods.[16] The analytical atomic spectroscopy literature is an excellent source for method and periodical reviews. These include a number of analytical journals, most of which are comprehensively reviewed on an annual basis.[17]

Computerised literature search data bases such as *Current Contents* and *Chemical Abstracts—Selected Topics* (Atomic Spectroscopy and Trace Elements) have proved very useful in attempting to keep up with the large amount of literature in this area.

2.2. General Laboratory Techniques

There are certain requirements in laboratory technique that are generally understood and practised to some extent by investigators determining 'trace' levels of inorganic elements in biological material. The quality of data generated by an investigator directly reflects the extent of these practices.[18]

The avoidance of contamination in trace analysis is critical.[19] Because of the low level of the trace inorganics in foods (which usually requires work at the low microgram or nanogram level) and the ubiquitous presence of significant levels of these elements in reagents, containers and air, the investigator must maintain an attitude of a 'useful paranoia' in evaluating and eliminating potential sources of contamination in every step of the entire study. These steps include initial collection of the bulk sample, storage and handling, homogenisation, analytical subsampling, analytical sample preparation, digestion or destruction of the organic matrix (solubilising the inorganics), and obtaining the analytical response in the appropriate instrumentation.

A most important aspect is the avoidance of airborne particulate contamination. The general laboratory area where any sample handling or transfer is carried out or where the sample is exposed to the atmosphere has to have control of airborne particulates. A minimum requirement is the availability of work stations within laminar-flow clean air hoods. In these stations the air is passed through a HEPA filter, which removes 99·97% of particulate matter greater than 0·3 μm in diameter. This produces an environment classified as Class 100, or less than 100 particles per cubic metre of air. To put this into perspective, an ordinary laboratory with no air filtration system may have a million or more particles per cubic metre. An excellent laboratory environment with laminar-flow hoods and scrupulous care for cleanliness may have an environment of Class 10 000 outside the immediate area of the laminar-flow work stations. With modern techniques capable of analysis at the nanogram or picogram level, a single particle of airborne particulate (usually inorganic in nature) falling into a sample can invalidate the results observed. A recent report[20] has extensively detailed design of suitable 'clean air' supplies for trace analyses.

More extensive discussion of general laboratory practices for trace element analyses has been reported elsewhere.[2,10,21] Discussions concerning appropriate sample preparation techniques can be found in these sources.

3. SPECIFIC TECHNIQUES: ATOMIC SPECTROMETRY

Of the many specific analytical techniques proposed or reported for use in the analysis of trace elements in food, the two which have been most utilised have been atomic absorption spectrometry (AAS) as a single-element technique and neutron activation analysis as a multi-element technique. The increasing interest in multi-element analysis and the commercial availability of instrumentation using the inductively coupled plasma (ICP) as an atomisation source has led to a large resurgence in interest in atomic emission spectrometry (AES) as a multi-element technique. ICP systems are being set up and utilised in several large-scale food analysis programmes,[22,23] and a considerable amount of food composition data will be generated by this method in the coming years. Other techniques that have been proposed and used for trace element analysis of foods include other atomic spectroscopic techniques, such as atomic fluorescence and atomic emission with either flame, DC plasma or arc source; spark source and stable isotope dilution mass spectrometry; x-ray techniques; electrochemical techniques; and chemical methods usually based on spectrophotometric determination of metal complexes formed in solution following a chemical separation of the element from the matrix.[9,10,12]

The following sections will review very briefly some of the basic principles and methods of the most widely used specific techniques.

3.1. Atomic Absorption Spectrometry

Modern atomic absorption spectrometry (AAS) developed from work of Walsh.[24] Beginning in 1960, it achieved rapid commercial success and today is the most popular method for metal determinations in general and the most widely used technique for analyses of trace elements in foods.[14] The popularity of AAS arises from its analytical specificity, good detection limits, excellent precision and relatively low cost. The main drawbacks of AAS have historically been the limited linear calibration range and the inability to analyse more than one element at a time. Solutions to both of these drawbacks have recently been reported, but have not yet been adopted in commercial instrumentation.[25,26]

The basic principles of AAS are well known and have been detailed in a number of textbooks and literature references to which the reader is referred. The first volume of this series includes a chapter on AAS in food analysis[21] which provides a general statement of the operating principles of AAS.

3.1.1. Instrumentation

A wide variety of AAS instruments is currently available. In general, each consists of a light source (a hollow cathode lamp, HCL), an atomisation source, and a dispersion/detection device. The most critical component, and the component for which the greatest analytical variability exists, is the atomiser. For AAS, three types of atomisers are commonly used: the flame, the graphite furnace, and chemical vaporisation (which will be discussed in Section 3.3).

(*a*) *Flame Atomisation*

Flame atomisation has been employed since the conception of AAS. A chemical flame is the simplest, oldest and best characterised of all the atomisers. The study of flame atomisation is a rather static field with no major breakthroughs having occurred in the last 10–15 years. During this time, flame atomisation has been used for the determination of almost every element in almost every conceivable sample matrix.

A wide variety of flames has been used for AAS, but the air–acetylene and nitrous oxide–acetylene mixtures are the most popular. Air–acetylene is the cooler of the two, about 2300 °C. It is more suitable for elements which ionise easily but, as a result, suffers more interferences for elements which tend to form stable compounds. The hotter nitrous oxide–acetylene flame, about 2700 °C, supplies the thermal energy needed to bring about dissociation of the compound-forming elements but is more difficult to use.

The use of HCLs for exact wavelength sources and the simplicity of the emission and absorption spectra result in flame AAS being a highly specific method. The detection limits range from 0·001 to 0·1 μg/ml and precisions of 0·1 to 0·3% can be attained. While the detection limits are very reasonable, they are approximately an order of magnitude worse than those for inductively coupled plasma or direct current plasma AES and two or three orders of magnitude worse than those for furnace AAS.

Interferences do occur for flame AAS, but they have been well characterised over the last 25 years. An extensive body of literature now exists for many elements analysed in a wide variety of materials. The analyst can almost always find a pertinent reference for every new sample type to be analysed.

Flame AAS is easy to operate. After the initial instrument set-up procedure the analyst need only aspirate the standards, samples and blanks in a systematic manner. Read-out of the sample signal is instantaneous, allowing the analyst to detect immediately any irregularities which might occur.

Flame AAS, in general, has proved readily adaptable to automation. Few modern spectrometers are available without a dedicated microprocessor. The analyst can now load the auto-sampler with standards, samples and blanks, in a systematic manner, start the operation and walk away. The spectrometer will measure each solution, calibrate, compute concentrations and print the results for each sample. A sequential multi-element instrument is commercially available that will repeat this automated determination for up to six elements without requiring the analyst to intervene. At present, no simultaneous multi-element AAS system is commercially available, although a variety of experimental systems has been reported. The present authors have demonstrated the use of continuum source atomic absorption with wavelength modulation for improved background correction to analyse a variety of food and beverage samples.[26,27] This system also utilises a novel computerised data reduction technique to extend greatly the dynamic measurement range.[25] This system has successfully been used for the simultaneous determination of up to 16 selected elements in reference materials and has been in routine use daily for analysis of food and biological materials. A commercially available simultaneous multi-element AAS system is likely to be seen in the future.

On the whole, flame AAS is the least expensive atomic spectrometric method. However, the cost of AA instruments covers a wide range depending upon the peripherals, computerisation and versatility desired. The simplest and least expensive instruments consist of the basic light source, flame atomiser, dispersion/detector configuration with electronics for computing absorbance, and a means of displaying the value. The most complex systems offer scanning monochromators, flame, furnace or ICP atomisation/excitation sources, HCL or electrodeless discharge lamp light sources, a dedicated microprocessor for spectrometer control, and a sophisticated data station for controlling automatic sampling, spectrometer operation, data processing and display. There are obviously many levels of complexity between these two extremes. The analyst must decide what capabilities and costs best meet the laboratory's needs.

(*b*) *Furnace Atomisation*

Furnace atomisation was first introduced by L'Vov in 1961.[28] The growth of the method of furnace atomisation has recently been quite rapid with new developments occurring at frequent intervals. There are indications that the study of furnace atomisation is just now approaching a period of stability, allowing for more complete characterisation of abilities and limitations.[29,30] Consequently, while furnace atomisation is presently a dynamic and exciting research field, it is simultaneously being employed for routine applications.

A furnace atomiser is quite simple in design. A small hollow carbon tube is positioned horizontally and an electrical current source is attached at both ends. A small opening is located in the middle of the tube which allows the solution to be injected into the tube. An electrical current passed through the tube produces resistive heating. Atomisation of the solution occurs in discrete steps. First, a current is passed through the tube which raises the temperature high enough to dry the sample by evaporating the solvent. Next, in the ashing or charring step, a higher current is employed to destroy the organic matter and evaporate the more volatile, undesirable inorganic components. Finally, a high current is applied to atomise the element of interest. Temperatures as high as 3000 °C can be reached in the atomisation step with a carbon furnace.

There are two methods of injecting a solution into the furnace. The most common approach is to pipette a discrete volume into a tube while at room temperature. The second method is to make an aerosol of the solution, similar to flame atomisation, and then blow a jet of the aerosol into the tube which has been heated above 100 °C. Sample volumes required range from 5 to 100 μl for discrete deposition and are comparable to flame atomisation for aerosol deposition.

The solution may be placed either directly on the inside wall of the furnace or on a 'platform'. A 'platform' was first suggested by L'Vov[31] and consists of a flat, or slightly curved, piece of carbon which bridges across the bottom of the furnace directly below, or opposite, the sample injection port. Material deposited directly on the furnace wall heats at the same rate as the wall and is atomised into an atmosphere whose temperature lags behind the wall temperature and which is changing rapidly with time. Material deposited on the platform is heated primarily by radiation from the furnace wall and therefore lags behind the wall temperature. When the delayed atomisation from the platform occurs, the temperature of the atmosphere in the furnace tube has had a chance to approach the wall temperature and is changing less rapidly. Platform atomisation into a more 'isothermal' atmosphere reduces many interferences that have been observed for furnace atomisation directly off the wall.

Furnace atomisation retains the specificity of flame atomisation and offers the best detection limits of any of the atomic spectrometry methods. Detection limits range from 0·01 to 1·0 ng/ml. Precision, however, is not as good, ranging from 1 to 4% for an optimised automated system, primarily because of inherent uncertainty of atomisation from the carbon surface.

It is currently acknowledged that interferences are worse for furnace atomisation than for flame atomisation. Since the entire sample is atomised into a small volume, non-specific background absorption and light refraction are much worse for furnace atomisation. In addition, severe matrix effects have also been reported. However, the use of platform atomisation, faster heating rates and matrix modification (chemical additions to the solution in the furnace which alter the atomisation characteristics of the analytical element and/or the interferent) have served to improve the accuracy of furnace determinations. A detailed discussion of interferences and a method of correction for known interferences has been presented elsewhere.[16] The importance of identifying and correcting for possible interferences in furnace analyses cannot be overstressed.

The literature on furnace atomisation is far from complete or conclusive. This is not surprising considering the stage of development of furnace atomisation. In most cases, the analyst must re-evaluate existing methods or develop new methods for each new sample matrix. Published literature may or may not be applicable to the furnace atomiser being used or the material being analysed.

Furnace atomisation is not easy to use for two major reasons: the inherent nature of the atomiser and the concentration levels being determined. The furnace atomiser requires several minutes between individual determinations and, because of its less precise nature, requires two or three determinations for each solution. Consequently furnace atomisation produces results at a slow rate and is not well suited for the determination of a large number of samples. The surface of the carbon tube steadily degrades as a function of the number of determinations. Coating the tube with pyrolytic carbon helps in slowing the degradation but does not cure the problem. Thus, furnace atomisation is subject to low-frequency noises of both a random and systematic nature.

Again, there is no commercially available multi-element furnace AAS. Preliminary data from the authors' laboratory suggest that multi-element furnace AAS can be accomplished and that the compromises for efficient simultaneous multi-element atomisation are less for furnace atomisation than for flame atomisation. However, significant modifications of presently available furnaces to prevent cross-contamination problems[32] and improvements in data acquisition methods to allow extended calibration ranges[25] are necessary before multi-element furnace AAS is likely to be developed commercially.

3.1.2. Applications

Both flame and furnace AAS are used for the determination of inorganic elements in food and beverages. The ease of operation and familiarity of flame atomisation make it the method of choice where large numbers of samples are to be analysed and when furnace detection limits are not required.

Flame atomisation has been used for the analysis of almost every metal and non-metal in the periodic table by direct and indirect methods. For methods for specific elements in foods, the reader is referred to several excellent sources. Koirtyohann and Pickett[5] and Crosby[4] provide general overviews for metal determinations in foods. Both devote more attention to the analysis of harmful metals than those of nutritional interest. Crosby presents a listing of analytical methods for the determination of As, Cd, Hg, Pb, Se and Sn and Koirtyohann covers Cd, Hg, Pb and Sn.

A more comprehensive survey of methodologies is presented by Fricke *et al.*[13] This survey is a fairly comprehensive bibliography, organised by element, for the determination of trace metals in food over the past 15 years. Each reference is listed with the food material analysed and with a brief summation of the ashing, separation/preconcentration and atomisation methods. A range of results is reported and recovery data are included, when available.

Ihnat[14] has written an extremely detailed chapter covering the analysis of foodstuffs by AAS. This work is essentially a 'how to' manual and covers all aspects of the physical sample preparation, sample treatment and the analytical process. Physical sample preparation and homogenisation methods are recommended for each type of foodstuff (cereals, dairy products, meat, fish, etc.). Sample digestion procedures, reagents, apparatus and standards preparation are discussed. Finally, general analytical protocol and specific techniques are recommended for 21 elements on an individual basis.

The maturity of flame AAS (it has been used for 25 years) means that methods for many metals in many materials have been tested and many of these methods are still current and applicable. A survey of the most recent AAS literature, as it applies to food and beverages, reveals mostly refinements of long-accepted sample preparation methods. Heanes,[33] Feinberg and Decauze[34] and Rowan *et al.*[35] used dry ashing to determine a variety of elements in food and plant materials. Agemian *et al.*,[36] Borriello and Sciandone,[37] Evans *et al.*[38] and Jackson *et al.*[39] explored wet digestion methods for the analysis of fish, beer and assorted foodstuffs. Each of these methods represents a fine tuning of existing methods to obtain more accurate results in the materials of interest.

Method development for furnace AAS is a much more dynamic field. This is a result of the improving equipment and the application of furnace atomisation to many areas of analysis for the first time. The current literature is of considerable interest to the analyst involved in furnace AAS. As mentioned previously, there are several sources of AAS reviews which permit papers on the determination of metals in foods to be located. These reviews are the *Analytical Chemistry* application reviews which are published every other year, the *Atomic Spectroscopy* bibliographies which appear every six months, and the *Annual Reports on Analytical Atomic Spectroscopy*.

In the last two years, furnace AAS has been used to determine trace metals in such foods as fish,[40,41] yeast and nutritional supplements,[42] edible oils,[43] orange juice,[44,45] oysters,[46b] legumes,[47] vegetarian products,[48] plants,[49] bovine liver[46a] and canned foods.[50] In many cases, these studies have originated primarily from instrumentation development goals.

3.2. Atomic Emission Spectrometry

Atomic emission spectrometry (AES) has experienced a rejuvenation with the development of the inductively coupled plasma (ICP) and the stabilised direct current plasma (DCP) as atomisation/excitation sources. With these sources, AES offers freedom from chemical interferences, precision comparable to flame AAS, detection limits superior to flame AAS (but not as good as furnace AAS), calibration ranges covering up to six orders of magnitude, and the ability to determine from 20 to 60 elements simultaneously. Emission spectra are, however, more complex than absorption spectra leading to greater possibility of spectral interferences.

3.2.1. Instrumentation

Conceptually, AES is a two-component system, consisting of an atomisation/excitation source and a dispersion/detection device. For quantitative analysis, the two most useful atomisation/excitation sources are the ICP and the DCP. Each of these will be discussed in more detail. Two distinct varieties of dispersion/detection devices are being used: the scanning spectrometer, which employs a single photomultiplier tube and scans rapidly through the wavelengths of interest, or the polychromator, which employs a separate fixed exit slit and photomultiplier tube for each wavelength of interest. The polychromator is preferred for routine determinations for large numbers of samples whereas a scanning monochromator is more versatile and is more useful for lighter sample loads where the elements of interest might vary. The resolution of the dispersion device is important since the high-energy excitation sources (the ICP and DCP) produce complex spectra. Resolving overlapping emission lines, or correcting for overlapping lines, is necessary for accurate determinations.

(*a*) *Inductively Coupled Plasma* (*ICP*)

The ICP was reported simultaneously in the mid-1960s by Wendt and Fassel[51] in the USA and by Greenfield *et al.*[52] in the UK. After a period of rapid growth in the early 1970s the ICP is now reaching a period of maturity. ICP-AES has been applied to almost every general type of sample matrix and its advantages and limitations have been well characterised.

The ICP torch consists of a series of concentric argon gas flows. The plasma gas flow is passed through an induction coil. The induction coil electromagnetically induces a radiofrequency argon plasma. A second flow, the carrier gas flow, lies inside and parallel to the plasma gas flow and injects the sample into the middle of the plasma. The carrier gas flow punches a hole through the middle of the plasma, giving the plasma a doughnut (toroid) shape.

A third argon flow, the coolant gas flow, surrounds the plasma gas and prevents the plasma from melting the surrounding quartz tube. The temperature of the argon ICP is approximately 10 000 °C. The atomised solution and carrier gas flow reach temperatures of 6000–7000 °C. Emission measurements are usually made 1–2 cm above the induction coil to obtain the maximum signal-to-noise ratio. Other gases and mini-torches have been used with varying degrees of success.

Two different designs of pneumatic nebulisers, the concentric nebuliser and the cross-flow nebuliser, are currently popular. These nebulisers are limited to a maximum of 1 % salt and perform more consistently at lower salt concentrations. A third nebulisation source, an ultrasonic nebuliser, based on a piezoelectric crystal vibrating at high frequencies (several million cycles per second), has been used but has not found widespread acceptance. The ICP, using pneumatic nebulisation, consumes solution volumes at a lower rate (1–2 ml/min) than flame AAS (5–8 ml/min) but with a nebulisation efficiency of only about 1 % compared to 5–10 % efficiency with flame AAS.

Several approaches to introducing micro-solutions have been proposed (for polychromatic detection systems) including funnels, micro-cups, flow injection and furnace atomisation.[53] The detection limits vary as a function of the method of introducing the sample and the sample volume used.

In general, the specificity of ICP-AES is not as good as that of either flame or furnace AAS because of the frequency of line overlap interferences. However, the organic matrix of food and beverages generally produces relatively simple spectra. Consequently, the specificity of ICP-AES is comparable to AAS for the determination of metals in food. The detection limits for ICP-AES are, on the average, better than flame AAS but worse than furnace AAS.[16] Detection limits for refractory elements are much better than those for flame AAS and comparable to furnace AAS. The precision of ICP-AES is comparable to or slightly worse than flame AAS.

Chemical interferences are essentially non-existent for ICP-AES but spectral interferences are quite common. Direct line overlap interferences have been extensively investigated in recent years and several spectral line atlases are now available.[54,55]

Mechanically, the operation of ICP-AES is quite similar to flame AAS. The nebuliser tube only needs to be moved from the blank to the standards and samples. However, ICP-AES lacks the single-element simplicity of flame AAS. The operator must interact with the computer to identify all data collected. It is not possible to follow each element simultaneously. In addition, the probability of cross-contamination is increased when calibration covers six orders of magnitude of concentration. As a result, time is required between atomisations to allow the nebuliser to clear. Thus, in many respects, ICP-AES is far more complex than flame AAS, more closely resembling furnace AAS with respect to the ease of operation.

Computerisation is necessary to utilise the full analytical capabilities of ICP-AES. Scanning monochromators and direct reading systems require computer operation in order to collect large amounts of data in short periods of time. The automation and computerisation features of ICP-AES should not be regarded as optional features, but instead as requirements for effective utilisation of the instrument.

ICP-AES is more expensive than flame or furnace AAS. The least expensive scanning ICP systems are roughly twice the cost of the most elaborate AAS instruments. An ICP source with a dedicated minicomputer and simultaneous detection capabilities for more than 60 elements costs four or five times as much as the scanning ICP systems. However, the cost aspect is not all negative. With this expenditure for equipment, and because of its more complex computerisation and electronics, laboratories generally employ only experienced analysts to operate the instrument. As a result, ICP results reported in the literature have generally been of a consistently high quality.

Finally, the real strength of ICP-AES is its multi-element capability. The large calibration range and the high energy source, free of chemical interferences, allow the same instrumental parameters to be used for all elements. There is no compromise in the detection limits for multi-element determinations.

(b) Direct Current Plasma (DCP)

The direct current plasma is a controlled DC arc with temperatures comparable to the ICP. Although a number of experimental systems have been reported, only one design is currently commercially available. This design employs three electrodes for stability with the resulting plasma looking like an inverted Y. The carrier gas originates from below the inverted Y and directs the nebulised material towards the junction of the two legs. Nebulisation is accomplished with a conventional flame AAS nebuliser. Emission measurements are made at a point just below the junction.

The DCP is sold with an echelle polychromator. The echelle polychromator offers resolution an order of magnitude better than most conventional monochromators and allows separation of many emission lines that cause overlapping interferences on other spectrometers. The echelle polychromator design allows the positioning of up to 20 end-on style photomultiplier tubes behind the focal plane.

The operational characteristics of DCP-AES are almost identical to ICP-AES. The difference lies in the introduction of the nebulised material into the plasma. For ICP-AES the carrier gas flows through the hole of the doughnut-shaped plasma, but for DCP-AES the carrier gas flows into the junction of the plasma legs and then around the plasma to either side. As a result, the nebulised material is exposed to lower temperatures in the DCP than in the ICP. Chemical interferences, which are virtually non-existent in the ICP, are more troublesome in the DCP, although still much less severe than they are for flame AAS. Since DCP-AES is only capable of determining 20 elements simultaneously, its costs lie towards the lower end of the range for ICP-AES instruments.

3.2.2. Applications

The ICP has not been applied to the determination of metals in foods to the same extent that flame AAS has. However, the applications which have been made are impressive in the number of elements simultaneously determined and the number of samples analysed.

The first extensive study of the determination of metals in biological materials by ICP was published by Dahlquist and Knoll.[56] This study looked at the determination of 19 elements in a variety of reference materials and botanical samples. More recently, H. Jones *et al.*,[57] J. Jones *et al.*[58] and Kuennen *et al.*[23] have used the ICP to determine from 14 to 25 elements in biological samples, foods, liquid protein and raw crop materials.

Two reviews on the application of ICP-AES to food samples have appeared recently. Jones and Boyer[15] reviewed the determination of metals in food by ICP-AES, and Mermet and Hubert[59] reviewed the application of ICP-AES to the analysis of biological materials. The latter review dealt primarily with the usefulness of ICP-AES in determining trace elements in biological materials. This same topic has been of concern to Barnes.[60] In general, ICP-AES does not have the detection limits to determine many trace level elements of biological interest. However, the usefulness of ICP-AES for large-scale food and plant material studies is obvious.

DCP-AES is not quite as popular as ICP-AES, but has been used effectively for the determination of metals in food and biological materials. Melton *et al.*[61] used DCP-AES for the determination of boron in plant material. DeBolt[62] measured 11 elements simultaneously in plant tissue. Woodis *et al.*[63] and Hunter *et al.*[64] analysed fertilisers. McHard *et al.*[65] used DCP-AES for the determination of 10 elements in orange juices.

3.3. Chemical Vaporisation

The chemical vaporisation technique has been widely used for atomic spectroscopic determinations of a number of metals that have generally proved difficult to atomise by more conventional means.

In chemical vaporisation the element of interest is chemically reduced to either a volatile covalent hydride (As, Bi, Ge, Pb, Sb, Se, Sn and Te) or the volatile metal (Hg). These volatile species are then introduced into the atomisation source of the spectrometer in the gas phase. This process generally leads to a greater sensitivity for these elements since all of the element in the sample is introduced to the atomisation source, whereas conventional nebulisation normally exhibits only 5–10% efficiency (at most) in aerosol formation. Hydride generation requires atomisation of the vapour by a thermal source to complete the generation of the atomic species, whereas the 'cold' vapour of Hg is suitable for direct determination. Hydride generation has been combined with atomisation by various chemical flames, electrically heated silica tubes, graphite furnaces, inductively coupled plasmas, microwave-induced plasma and DC plasmas.

In general, the chemical reduction step is specifically designed for the element of interest and the sample type. Interferences for the chemical treatment step are common and highly specific to the element, the method of reduction and the sample matrix. There is little standardisation of equipment and methods. Only in recent years has equipment specifically designed for vapour generation appeared on the market. Automation of the vapour generation process has led to improved accuracies and precisions.

A general review of chemical vaporisation has been published by Godden and Thomerson.[66] Critical appraisals of methods have been presented by Snook[67] and Ihnat and Thompson.[68] The latter article is of particular interest. After extensive statistical evaluation, the authors concluded that chemical vaporisation was unsuitable as a reference method owing to the lack of standardisation of the basic method.

Chemical vaporisation has been used for the determination of As, Bi, Ge, Hg, Pb, Sb, Se, Sn and Te in a wide variety of matrices.[69] The specificity of chemical vaporisation has generally made it a single-element method. However, Wolnick *et al.*[70] demonstrated that multi-element determinations could be obtained using an ICP. The same group obtained multi-element detection limits an order of magnitude better using a cryogenic trap.[71]

4. SPECIFIC TECHNIQUES: OTHER

4.1. Neutron Activation Analysis

Neutron activation analysis (NAA) has been used for the analysis of the elemental content of a wide variety of biological samples. The principles, instrumentation and applications of NAA to biological samples have been reviewed.[72]

This technique can be very sensitive for a wide range of elements and has the advantage of multi-element analysis. There are relatively few interference and matrix effects in NAA and these are fairly well documented. This technique can supply accurate, highly precise results for most of the trace elements of biological interest.

The major drawback with this technique is that appropriate facilities and equipment have to be rather elaborate, i.e. a source of neutrons and radiochemistry facilities have to be available.

The technique does not lend itself readily to the analysis of a large number of samples in a routine mode, and has been utilised mainly as a research tool for food and biological material analysis. Applications of NAA for determination of trace elements in foods have included multi-element estimation of dietary intake[73] and use of NAA for analysis of 'total diets'.[74] NAA is particularly useful for elements such as iodine[75] and mercury[76] which are very difficult to analyse by other techniques. One particular problem with analysing foodstuffs and other biological materials which contain much sodium, potassium, phosphorus and bromine is the very high activity of the radionuclides of these elements. This activity dominates the gamma spectrum and makes determination of most other elements very difficult or impossible. Either chemical separation or purely instrumental technique must be used after a longer decay period. A remote-controlled system for NAA multi-element determination in foodstuffs has been developed and carefully validated.[77]

4.2. Mass Spectrometry

A number of mass spectrometric techniques have been utilised to determine trace elements in specific biological materials.[78] Spark source mass spectrometry (SSMS) is particularly useful for simultaneous multi-element trace analysis for survey purposes. It can be used to determine over 30 elements with good sensitivity and accuracy and precision on the order of 10% or better. At the present time, SSMS, though it may be applicable if facilities and expertise already exist, would not be a method of choice if one were setting up a new programme to survey trace element content in foods. The limited number of elements of biological interest which can be determined using SSMS and the relatively poor precision make it a much less favourable choice than most of the atomic spectroscopic techniques.

Isotope dilution mass spectrometry (IDMS) has seen primary use in analysis of selected trace elements for analyses requiring very high accuracy and precision in the order of 0·1 %. This technique has been particularly useful for the National Bureau of Standards for the certification of biological Standard Reference Materials.[7] IDMS may be the ultimate technique currently available in terms of sensitivity, accuracy and precision. However, as currently utilised it is primarily a single-element technique requiring chemical separation and extreme caution to avoid contamination. The analysis time is somewhat lengthy, and instruments have usually been specially built and are therefore quite expensive. Utilisation of volatile metal chelates to introduce the trace element into the mass spectrometer may lead to increased use of IDMS for more routine analysis for several trace elements in biological materials.[79] The use of stable IDMS for trace element determination in biological systems has been reviewed.[80]

5. QUALITY ASSURANCE FOR TRACE ELEMENT ANALYSIS

The most important aspects of any analytical method are the steps taken to ensure the quality of the results. Quality assurance is the system of activities whose purpose is to provide assurance that the overall quality control limits are being met.[81]

Proper quality assurance stems from an understanding of the entire analytical system, from sampling through the analytical aspects of sample treatment and measurement to data handling and evaluation. The analytical methodology must be fully understood and validated. Expected accuracy and precision of the result must be established. Defined quality control procedures must be developed and employed during use of the methodology.

With the recent advances in modern analytical techniques the need for quality assurance is even more critical. The current trend towards multi-element determinations, automation and computerisation has resulted in the analyst being less directly involved in the actual analytical measurement process. Deviations from normal operation for automated, computerised procedures, which previously were noted and corrected by the analyst as a part of operator interaction throughout the procedure, must now be detected and corrected by an appropriate computer algorithm. All too frequently the analyst accepts computerised results without question. This point is well made by Frank and Kowalski[82] in their review on chemometrics, which should be required reading for everyone involved with automated and computerised methods. Without the proper quality assurance, the latest advances in analytical methodology result only in erroneous results being turned out at a faster rate.[83]

5.1. Sampling

Sampling is the selection, from a population, of a finite number of individuals to be analysed. The object is to extrapolate the results obtained for the analysis of the individuals to characterise the entire population. A bias in sampling can produce a bias in the characterisation of the population. Proper sampling techniques must be used to ensure that the individuals are representative of the whole.

The problems in establishing a valid sampling of foods are considerable because of the many variables. For processed foods the analyst must consider the wide variety of brand names available, geographical location, seasonal variation and the effect of shelf life. There are also longer-term variations associated with the development of new processing methods and new sources of raw materials. The number of variables associated with fresh foods and commercially prepared foods are even greater. Unless valid statistical methods are used, the results of the study may not be representative beyond the limited population analysed.

Statistically valid sampling is a complex science. The reader is referred to the many excellent publications on the topic for detailed discussions in areas of interest.[6,84,85]

5.2. Validation of Analytical Data

To develop a meaningful routine method of analysis, the whole procedure, including sample preparation and handling, must be fully understood, critically examined, and validated. The strengths, weaknesses, limitations and ruggedness of the analytical aspects of a procedure must be documented by appropriate research and development. Potential matrix effects must be identified for each type of sample. Each type of food is basically a different chemical matrix. For a particular nutrient, a method that is valid for one type of food may not be valid for another. Certain aspects of the instrumentation and/or sample preparation might cause significant differences for different matrices. Such effects must be identified and corrected in the development of an accurate procedure.

Once the analytical aspects of the procedure are well known, and the method is under control (i.e. reproducible results with high precision can be obtained), the method must be validated to give correct or accurate results of the 'true' value of the analyte in the samples.

After a method has been developed, verified and established as routine, it must be monitored by appropriate procedures that are established to ensure that the data obtained during the use of the method remain valid. The key to this quality control (QC) is the availability and use of appropriate control samples.

5.2.1. Method Validation

In the course of establishing appropriate procedures for analysing trace elements in foods, the methods must first be brought under control, i.e. capable of reproducing the analysis with acceptable precision (usually $<10\%$ coefficient of variation). The method must next be validated, i.e. shown to give the right or 'true' value. A rigorous quality control procedure must then be incorporated into the method to assure control and accuracy on a day-to-day basis.

Precision can be obtained by identifying and controlling the significant sources of random error, both instrumental and sample handling. Method validation can be obtained by several different approaches, including confirmation of analysis by independent methods, or correct analysis of certified reference materials. Quality control procedures require establishment of appropriate check samples to be run on a batch-to-batch basis.

(*a*) *Two Independent Methods*

The establishment of independent methods for determining the same elements is the ideal method for validating analytical results. Independent methods, each based on a different physical principle, seldom suffer from the same systematic biases, or interferences. If different sample preparation methods are employed, then analytical agreement of the two methods allows one to place a great deal of confidence in the result. This approach is not usually considered since it is either beyond the capability of most laboratories or it is not economically feasible. This is especially true for the analysis of trace elements. The development of a single state-of-the-art method can be quite draining of time, money and resources. Development of a second method is not usually considered unless the purpose is to produce highly reliable values for an appropriate reference material.

(*b*) *Reference Materials*

The most common means of validating results is to obtain accurate determinations for a certified reference material of the same chemical composition.

Certified reference materials are reference materials accompanied by a certificate issued by a recognised official standards agency. These certificates give the reference value of the component plus confidence limits. The National Bureau of Standards (NBS) is the US source of certified reference materials. The NBS Standard Reference Materials (SRMs) are carefully prepared for homogeneity and stability and are characterised by at least two independent analytical methods.[7] NBS currently lists about 10 biological SRMs which might have some application for foods.

The International Atomic Energy Agency (IAEA) also has several biological materials certified for a number of inorganic elements including trace elements.[86] Several other laboratories have issued reference materials and a directory of these materials is available.[87] There is promise for several new materials in the near future.[88,89]

The major problem associated with reference materials is finding one similar to the samples to be analysed. The limited number of currently available food reference materials means that, in many cases, a reference material must be used which only superficially resembles the sample. Cluster analysis procedures have been used to show that, for metals, there is relatively little composition overlap between the available biological reference materials (from NBS and the IAEA) and 160 of the most commonly consumed foods in the USA.[90] In addition, the range of concentrations of the elements in the reference materials is not fully compatible with the range of concentrations found in foods. There is currently a great need for a wider range (both of food types and of elemental concentrations) of food reference materials.

(*c*) *Other Validation Methods*

In the absence of a second, independent method or a suitable reference material, results can be compared to previously reported values in the literature. Agreement with the published values lends strong support to the accuracy of the method, but disagreement can be very unsatisfying. In most cases, there is no way to resolve differences between the data without the development of an independent method. It is also possible that the analytical differences arise from the samples being different. This would be best resolved by an exchange of samples.

Analysis of a common exchange sample by a group of laboratories is another means of evaluating analytical results. Agreement with other laboratories lends strong support to the validity of the method, but differences may be difficult to resolve. If the group exchanging samples is not sufficiently large, or if everyone employs the same analytical method, the consensus result may not be the most accurate value. The agreement between laboratories is more significant if more than one independent method is employed.

5.2.2. Quality Control Samples

Primary or certified reference materials can be expensive and are not available in large quantities. These materials are generally used only for the initial validation of a method. For routine usage, it is necessary for laboratories to develop secondary reference materials or quality control (QC) samples. The QC samples usually consist of a particularly large amount of an individual sample or several samples pooled together to be more representative of the samples to be analysed. The QC samples are carefully characterised using certified materials (if possible) and the best possible quality assurance methods to determine the 'true' value of their components and the normal ranges of variation.

The QC samples are used in a number of ways. They may be analysed in conjunction with the samples in a random or systematic pattern. For atomic spectrometric methods, where the instrument must be recalibrated at frequent intervals, it is desirable to analyse at least one QC sample for each calibration. QC samples may be labelled or analysed 'blind' (unidentified to the analyst). The results of the QC sample are compared with the predetermined 'true' values and ranges of variation. Poor accuracy or precision of a QC sample casts doubt on the validity of the results of all the samples analysed at that time, and usually leads to rejection of the analyses for that group of samples. Rejection criteria must be established by the analyst ahead of the time of actual analysis.

6. DATA HANDLING AND EVALUATION

After an analysis, the raw data must be used to convert the sample readings into concentrations. Thus, a calibration function must be determined which best fits the calibration standards, the sample concentrations must be computed, and the results must be evaluated with respect to their reliability and with respect to the experimental design. The recent boom in computer technology has had a dramatic impact on all aspects of data handling and evaluation.

6.1. Calibration Function

The historical approach to calibration was to restrict the analysis to th linear range, plot the calibration standards on graph paper, throw ou those points which didn't look 'right', draw the best straight line to fit th remaining points, and then convert the sample signals to concentrations using the 'linear calibration'. With a computer, much more sophisticated mathematical operations are possible. Complex equations can now be used for calibration functions and statistical methods of analysis can be applied to the raw data and the computed concentrations.

The first step in any calibration procedure is to inspect the calibration data. The standards must show good agreement between repeat determinations. A systematic change in the standard values usually indicates drift of one of the analytical parameters. The standards must also have an acceptable signal-to-noise ratio. Standards whose signals have a poor signal-to-noise ratio can deviate considerably from the true value unless determined a sufficient number of times to obtain an accurate mean value. And finally, the standard signals must progress logically in the order and proportion of their concentrations. Although a standard may appear in error by visual inspection, elimination on the basis of a statistical test may prove difficult. Many times the best approach is to redetermine the standard or make up a new standard.

The most accurate means of calibration, in either the absorption or emission mode, is to use a pair of standards which bracket the sample as tightly as possible. This approach eliminates any concern about linearity and reduces the time interval between the determination of the standards and the sample, minimising fluctuation, or drift, errors. Of course this approach is not practical for large numbers of sample, when the samples include a wide range of concentrations, or when operating in the multi-element mode.

6.2. Data Evaluation

6.2.1. Uncertainty of the Data

Computation of sample concentrations is never the final step in the investigative process. Even though the analyst may have no further involvement in the project, the results will always be used (or misused) by someone. All too often, the end users of the data lack an analytical background and tend to interpret the results as absolute, forgetting the inherent uncertainty in each value. It is therefore the duty of the analyst to assign a value of uncertainty to the data in order to ensure its proper usage.

Repeated determinations of the same sample can be used to compute a standard deviation which reflects the measurement error. The significance of the standard deviation can be expanded by including more variables. If the repeat determinations include different preparations of the sample, then the standard deviation will reflect measurement and sample preparation uncertainty. By performing the repeat measurements on different days, the standard deviation will reflect the uncertainties listed above as well as those uncertainties arising from daily variations in instrument performance and calibration. The best method of characterising analytical errors is to include multiple determinations made on different days by different analysts of different sample preparations. For this reason, the standard deviations of the quality control sample are usually the most accurate measurement of analytical uncertainty.

6.2.2. Statistical Analysis

Depending on the purpose of the study and the sampling scheme employed, the data may be tested using a wide variety of statistical methods. Some of the simpler tests are: the *t*-test to determine whether data or groups of data are significantly different, the *F*-test to determine whether the variances of the data are different, one-way and two-way analysis of variance to test for one- and two-dimensional patterns in the data, and multiple regression analysis to determine whether the data are dependent on preselected variables. These tests are described in most statistics texts. In addition, statistics packages are available for almost every micro- and minicomputer and for every large computer.

With the current emphasis on multi-element determinations, automation and computerisation, it is now possible to generate data faster than ever before. The weakest link in the analytical process is increasingly becoming the interpretation of the data. Development of the field of study of 'chemometrics' has offered great potential for strengthening this data evaluation link. This discipline uses mathematical and statistical methods to design an optimal measurement procedure and to provide maximum chemical information by analysing chemical data.[82] A number of chemometric approaches and tools are available for evaluation of multicomponent data, such as that generated in multi-element atomic spectroscopic studies. The techniques can be very useful in determining and sorting out multi-factorial relationships within groups of samples. These chemometric concepts are only recently being applied to the development of automated, computerised, analytical atomic spectroscopic systems to generate trace element data in foods.

6.3. Computerisation

Data which previously kept the analyst busy for weeks can now be processed in a matter of minutes using computerised mathematical approaches which previously were too complex to be used. However, computerisation does create additional quality assurance problems for the analyst. Because a computer does not have the flexibility of the human mind, deviations from the expected data patterns can lead to erroneous results. This is not because of inaccuracies in the number handling of the computer but because of the rigid nature of the computer logic. Once established, a program will treat each piece of data in an identical fashion. Whereas the analyst might note that one of the standards in a calibration was reading slightly high or low, the computer, unless specifically programmed to check the standards (which would be difficult), will accept the data as absolute. This problem is compounded in multi-element spectrometers where most of the collected data have never been seen by the analyst. Unless the computer is programmed to detect a wide variety of errors and to evaluate the validity of the data, significant analytical errors can result.

Every step in the data reduction process, which was previously performed by the analyst, must be translated into a computer algorithm, or a series of algorithms. This is an extremely complex undertaking. The more completely a program mimics the thought processes of the analyst, the fewer logical errors are to be expected. However, few programs achieve such a level of sophistication. As programs become more versatile, they become more complex and cumbersome. The logic of more complex programs becomes difficult to follow and more susceptible to errors. On the other hand, conceptually simple programs are often too narrow in their application. As a result, most programs are a compromise: logically as simple as possible but able to handle the most common problems.

It is imperative that the analyst be familiar with how the data are handled by a computer program. This knowledge allows the analyst to inspect the data and results, detect errors, and anticipate conditions for which the program algorithms are not appropriate. Obtaining listings of commercially obtained programs is not always possible. Most manufacturers do not make original programs, or 'source' listings, available to the users. Thus, the analyst is able to obtain only a general idea of how the data are being handled. Certainly, the analyst must try to evaluate the program's performance under as many different circumstances as possible. In most cases, analyses of reference materials and quality control samples are the best methods for evaluating the accuracy of the analytical methods and the computer programs.

7. CONCLUSION

Many new tools and advanced methods in instrumentation, computerisation and automation are available in the analytical community for the analysis of trace elements. Most analysts have not taken full advantage of these advances to generate needed data on food composition. In order to apply these advances effectively for routine determinations of trace elements in foods, we must critically test and validate procedures of sample preparation and properly define the analytical methodology and quality control procedures. Establishment of these quality control procedures will depend on the further development of appropriate standards and food reference materials with certified trace element content.

REFERENCES

1. Stewart, K. K. (1980). In *Nutrient Analysis of Foods: The State of the Art for Routine Analysis*, Stewart, K. K. (Ed.), Association of Official Analytical Chemists, Washington, DC, p. 1.
2. Wolf, W. R. (1982). In *Clinical, Biochemical and Nutritional Aspects of Trace Elements*, Prasad, A. (Ed.), Alan R. Liss, New York, p. 427.
3. Wolf, W. R. (1981). In *Environmental Speciation and Monitoring Needs for Trace Metal-Containing Substances from Energy-Related Processes*, National Bureau of Standards Special Publication 618, Bruschman, F. E. and Fish, R. E. (Eds), Proc. Workshop Environmental Speciation, Gaithersburg, MD, 1981, US Government Printing Office, Washington, DC, p. 235.
4. Crosby, N. T. (1977). *Analyst*, **102**, 225.
5. Koirtyohann, S. R. and Pickett, E. E. (1975). In *Flame Emission and Atomic Absorption Spectrometry*, Vol. 3: *Elements and Matrices*, Dean, J. A. and Rains, T. C. (Eds), Marcel Dekker, New York, Chap. 17.
6. NBS (1976). *Accuracy in Trace Analysis: Sampling, Sample Handling Analysis*, Vols. 1 and 2, National Bureau of Standards Special Publication 422, LaFleur, P. D. (Ed.), Proc. 7th Material Research Symposium, Oct. 1974, US Government Printing Office, Washington, DC.
7. NBS (1977). *Procedures Used at the National Bureau of Standards to Determine Selected Trace Elements in Biological and Botanical Materials*, National Bureau of Standards Special Publication 492, Mavrodineanu, R. (Ed.), US Government Printing Office, Washington, DC.
8. Risby, T. T. (Ed.) (1979). *Ultratrace Metal Analysis in Biological Sciences and Environment*, Advances in Chemistry Series No. 172, American Chemical Society, Washington, DC.
9. Berman, E. (1980). *Toxic Metals and Their Analysis*, Heyden, London.
10. IAEA (1980). *Elemental Analysis of Biological Materials: Current Problems and Techniques with Special Reference to Trace Elements*, Technical Reports Series No. 197, International Atomic Energy Agency, Vienna.
11. Bratter, P. and Schramel, P. (Eds) (1980). *Trace Element Analytical Chemistry in Medicine and Biology*, Proc. Workshop, Neuhenberg, 1980, W. deGruyter, Berlin.

12. VEILLON, C. and VALLEE, B. L. (1978). Atomic spectroscopy in metal analysis of enzymes and other biological materials. In *Methods in Enzymology*, Fleicher, S. and Packer, L. (Eds), Academic Press, New York, Vol. 54(25), p. 446.
13. FRICKE, F. L., ROBBINS, W. B. and CARUSO, J. A. (1979). *Progr. Anal. At. Spectrosc.*, **2**, 185.
14. IHNAT, M. (1981). Application of atomic absorption spectrometry to the analysis of foodstuffs. In *Atomic Absorption Spectrometry*, Cantle, J. E. (Ed.), Vol. 5 of *Techniques and Instrumentation in Analytical Chemistry*, Elsevier, Amsterdam, p. 139.
15. JONES, J. W. and BOYER, K. W. (1979). *Applications in Inductively Coupled Plasma Emission Spectroscopy*, Barnes, R. M. (Ed.), Franklin Institute Press, Philadelphia, p. 83.
16. HARNLY, J. M. and WOLF, W. R. (1984). Atomic spectrometry for inorganic elements in foods. In *Instrumental Analysis of Foods and Beverages: Modern Techniques*, Charalambous, G. (Ed.), Academic Press, New York, p. 451.
17. *Annual Reports* (1971–1981) *on Analytical Atomic Spectroscopy*, The Chemical Society, Burlington House, London W1V 0BN, Vols. 1–10.
18. ACS (1980). Guidelines for data acquisition and data quality evaluation in environmental chemistry. *Anal. Chem.*, **52**, 2242.
19. ZEIF, M. and MITCHELL, J. W. (1976). *Contamination Control in Trace Element Analysis*, Wiley-Interscience, New York.
20. MOODY, J. R. (1982). *Anal. Chem.*, **54**, 1358A.
21. COWLEY, K. M. (1978). Atomic absorption spectrometry in food analysis. In *Developments in Food Analysis Techniques—1*, King, R. D. (Ed.), Applied Science Publishers, London, p. 293.
22. BOYER, K. W., TANNER, J. T. and GAJAN, R. J. (1978). *Am. Lab.*, **10**(2), 51.
23. KUENNEN, R. W., WOLNICK, K. A., FRICKE, F. L. and CARUSO, J. A. (1982). *Anal. Chem.*, **54**, 2146.
24. WALSH, A. (1955). *Spectrochim. Acta*, **7**, 108.
25. HARNLY, J. M. and O'HAVER, T. C. (1981). *Anal. Chem.*, **53**, 1291.
26. HARNLY, J. M., O'HAVER, T. C., GOLDEN, B. M. and WOLF, W. R. (1979). *Anal. Chem.*, **51**, 2007.
27. HARNLY, J. M. and WOLF, W. R. (1981). 15th Middle Atlantic Region Meeting, ACS, Jan. 1981, Paper No. 29.
28. L'VOV, B. V. (1961). *Spectrochim. Acta*, **17**, 761.
29. KOIRTYOHANN, S. R. and KAISER, M. (1982). *Anal. Chem.*, **54**, 1515A.
30. SLAVIN, W. (1982). *Anal. Chem.*, **54**, 685A.
31. L'VOV, B. V. (1976). 3rd Annual Meeting of the Federation of Analytical Chemistry and Spectroscopy Societies, Philadelphia, Report No. 214.
32. HARNLY, J. M. (1982). 9th Annual Meeting of the Federation of Analytical Chemistry and Spectroscopy Societies, Philadelphia, Paper No. 297.
33. HEANES, D. C. (1981). *Analyst*, **106**, 182.
34. FEINBERG, M. and DUCAUZE, C. (1980). *Anal. Chem.*, **52**, 207.
35. ROWEN, C. A., ZAJICEK, O. T. and CALABRESE, E. J. (1982). *Anal. Chem.*, **54**, 149.
36. AGEMIAN, H., STURTEVANT, D. P. and ANSTEN, K. D. (1980). *Analyst*, **105**, 125.
37. BORRIELLO, R. and SCIANDONE, G. (1980). *At. Spectrosc.*, **1**, 131.

38. EVANS, W. H., DELLAN, D., LUCAS, B. E., JACKSON, F. J. and READ, J. (1980). *Analyst*, **105**, 529.
39. JACKSON, F. J., READ, J. T. and LUCAS, B. E. (1980). *Analyst*, **105**, 359.
40. MAY, T. W. and BRUMBAUGH, W. G. (1982). *Anal. Chem.*, **54**, 1032.
41. LANGMYHR, F. J. and ORRE, S. (1980). *Anal. Chim. Acta*, **118**, 307.
42. THOMPSON, D. D. and ALLEN, R. J. (1981). *At. Spectrosc.*, **2**, 53.
43. SLIKKERVEER, F. J., BRAOD, A. A. and HENDRIKSE, P. W. (1980). *At. Spectrosc.*, **1**, 30.
44. NAGY, S. and ROUSSEFF, R. L. (1981). *J. Agric. Fd Chem.*, **29**, 889.
45. HARBACH, D., DIEHL, H., TIMIS, J. and HUNTEMAN, D. (1980). *Fresenius Z. Anal. Chem.*, **301**, 215.
46. (a) CHAKRABARTI, C. L., WAN, C. C. and LI, W. C. (1980). *Spectrochim. Acta*, **35B**, 93; (b) *ibid.*, **35B**, 547.
47. ROCKLAND, L. B., WOLF, W. R., HAHN, D. M. and YOUNG, R. (1979). *J. Fd Sci.*, **44**, 1711.
48. FREELAND-GRAVES, J. H., EBANGIT, M. L. and BODZE, P. W. (1980). *J. Am. Diet. Ass.*, **77**, 648.
49. DUPRIE, S. and HOENIG, M. (1980). *Analysis*, **8**, 153.
50. DAKEBA, R. W. (1979). *Anal. Chem.*, **51**, 902.
51. WENDT, R. H. and FASSEL, V. A. (1964). *Anal. Chem.*, **37**, 920.
52. GREENFIELD, S., JONES, T. L. and BERRY, C. T. (1964). *Analyst*, **89**, 713.
53. BOYKO, W. J., KEBBER, N. K. and PATTERSON III, J. M. (1982). *Anal. Chem.*, **54**, 188R.
54. BOUMANS, P. W. J. M. (1980). *Line Coincidence Tables for Inductively Coupled Plasma Atomic Emission Spectrometry*, Pergamon Press, New York.
55. PARSONS, M. L. and FORSTER, A. (1980). *An Atlas of Spectral Interferences in ICP Spectroscopy*, Plenum Press, New York.
56. DAHLQUIST, R. L. and KNOLL, J. W. (1978). *Appl. Spectrosc.*, **32**, 1.
57. JONES, H. O. L., JACOBS, R. M., FRY Jr, B. E., JONES, J. W. and GOULD, J. H. (1980). *Am. J. Clin. Nutr.*, **32**, 2545.
58. JONES, J. W., CAPAR, S. and O'HAVER, T. C. (1982). *Analyst*, **107**, 353.
59. MERMET, J. M. and HUBERT, J. (1982). *Progr. Anal. At. Spectrosc.*, **5**, 1.
60. BARNES, R. M. (1982). 9th Annual Meeting of the Federation of Analytical Chemistry and Spectroscopy Societies, Philadelphia, Paper No. 62.
61. MELTON, J. R., HOOVER, W. L., MORRIS, P. A. and GERALD, J. A. (1978). *J. Ass. Off. Anal. Chem.*, **61**, 504.
62. DEBOLT, D. C. (1980). *J. Ass. Off. Anal. Chem.*, **63**, 802.
63. WOODIS Jr, T. C., HOLMES Jr, J. H., ARDIS, J. P. and JOHNSON, F. J. (1980). *J. Ass. Off. Anal. Chem.*, **63**, 1245.
64. HUNTER, G. B., WOODIS Jr, T. C. and JOHNSON, F. J. (1981). *J. Ass. Off. Anal. Chem.*, **64**, 25.
65. MCHARD, J. A., FOULK, S. J., NIKDEL, S., ULLMAN, A. H., POLLARD, B. D. and WINEFORDNER, J. D. (1979). *Anal. Chem.*, **51**, 1613.
66. GODDEN, R. G. and THOMERSON, D. R. (1980). *Analyst*, **105**, 1137.
67. SNOOK, R. D. (1981). *Anal. Proc.*, **18**, 342.
68. IHNAT, M. and THOMPSON, B. K. (1980). *J. Ass. Off. Anal. Chem.*, **63**, 814.
69. HORLICK, G. (1982). *Anal. Chem.*, **54**, 276R.

70. Wolnick, K. A., Fricke, F. L., Hahn, M. H. and Caruso, J. A. (1981). *Anal. Chem.*, **53**, 1030.
71. Hahn, M. H., Wolnick, K. A., Fricke, F. L. and Caruso, J. A. (1982). *Anal. Chem.*, **54**, 1048.
72. Guinn, V. P. and Hoste, J. (1980). In Technical Report No. 197, IAEA, Vienna, p. 105.
73. Schelenz, R. (1977). *J. Radioanal. Chem.*, **37**, 539.
74. Tanner, J. T. and Friedman, M. H. (1977). *J. Radioanal. Chem.*, **37**, 529.
75. Allegrini, M., Boyer, K. W. and Tanner, J. T. (1981). *J. Ass. Off. Anal. Chem.*, **64**, 1111.
76. Tanner, J. T. and Forbes, W. S. (1975). *Anal. Chim. Acta*, **74**, 17.
77. Schelenz, R. and Diehl, J.-F. (1976). In NBS Special Publication No. 422, p. 1173.
78. Morrison, G. (1980). In Technical Report No. 197, IAEA, Vienna, p. 201.
79. Veillon, C., Wolf, W. R. and Guthrie, B. E. (1979). *Anal. Chem.*, **51**, 1022.
80. Veillon, C. and Alvarez, R. (1983). In *Metal Ions in Biological Systems*, Sigel, H. (Ed.), Marcel Dekker, New York, Vol. 16, p. 103.
81. Taylor, J. K. (1981). *Anal. Chem.*, **53**, 1588A.
82. Frank, I. E. and Kowalski, B. R. (1982). *Anal. Chem.*, **54**, 232R.
83. Harnly, J. M. and Wolf, W. R. (1984). Quality assurance for atomic spectroscopy. In *Instrumental Analysis of Foods and Beverages: Modern Techniques*, Charalambous, G. (Ed.), Academic Press, New York, p. 483.
84. Youden, W. J. and Steiner, E. H. (1967). *Statistical Manual of the Association of Official Analytical Chemists*, AOAC, Washington, DC.
85. Kratochvil, B. and Taylor, J. K. (1981). *Anal. Chem.*, **53**, 924A.
86. Parr, R. M. (1980). In *Trace Element Analytical Chemistry in Medicine and Biology*, Bratter, P. and Schramel, P. (Eds), W. de Gruyter, Berlin, p. 631.
87. ISO (1982). *Directory of Certified Reference Materials*, International Standardisation Organisation/Remco, Geneva, Switzerland.
88. Ihnat, M., Cloutier, R. A. and Wolf, W. R. (1982). 9th Annual Meeting, Federation of Analytical Chemistry and Spectroscopy Societies, Philadelphia (abstract).
89. Muntau, H. (1980). In *Trace Element Analytical Chemistry in Medicine and Biology*, Bratter, P. and Schramel, P. (Eds), W. de Gruyter, Berlin, p. 707.
90. Wolf, W. R. (1982). Evaluation of available reference materials for potential use in analysis of foods. Eastern Analytical Symposium, New York, Nov. 1982 (abstract).

Chapter 3

DETERMINATION OF MYCOTOXINS

H. P. VAN EGMOND
National Institute of Public Health,
Bilthoven, The Netherlands

1. INTRODUCTION

Mycotoxins may be defined as metabolites of fungi which evoke pathological changes in man and animals. The term 'mycotoxin' is derived from the Greek words *mykes* (fungus) and *toksikon* (poison). Mycotoxicoses may be defined as the toxicity syndromes resulting from the intake of mycotoxins by man and animals, usually by ingestion.

The diseases caused by mycotoxins have been known for a long time. The first recognised mycotoxicosis was probably ergotism,[1] a disease characterised by necrosis and gangrene and better known in the Middle Ages under the name 'holy fire', which was caused by the intake of grain contaminated with sclerotia of *Claviceps purpurea*.

Another mycotoxicosis recognised to have seriously affected human populations is alimentary toxic aleukia (ATA).[2,3] The symptoms in man take on many aspects, including leukopenia, necrotic lesions of the oral cavity, the oesophagus and stomach, sepsis, haemorrhagic diathesis and exhaustion of the bone marrow. The disease was induced by eating overwintered mouldy grain and occurred in many areas in Russia, especially during World War II. It has been reported[4] that in 1944 more than 10% of the population in certain districts of Russia was affected and many fatalities occurred. The fungi responsible for these accidents belong to the genera *Fusarium* and *Cladosporium*.

In Japan, toxicity associated with yellow-coloured mouldy rice has been a problem, especially after World War II, when rice had to be imported from various countries.[5] The intake of 'yellow rice' by man caused vomiting, convulsions and ascending paralysis. Death could also occur within 1–3 days after the first signs of the disease appeared. The toxin-producing fungi in yellow rice belong to the genus *Penicillium*.

Despite the forementioned examples of mycotoxin-caused diseases in man, mycotoxicoses remained the 'neglected diseases'[6] until the early 1960s, when this attitude changed drastically owing to the outbreak of Turkey X Disease in the UK.[7] Within a few months more than 100 000 turkeys died, mainly in East Anglia and southern England. In addition, the death of thousands of ducklings and young pheasants was reported.[8] The appearance of Turkey X Disease led to a multidisciplinary approach to investigate the cause of the problem. These efforts were successful and the cause of the disease was traced to a toxic factor occurring in the Brazilian groundnut meal which was used as a protein source in the feed of the affected poultry. The toxic factor seemed to be produced by two fungi, *Aspergillus parasiticus* and *Aspergillus flavus*, and hence the name 'aflatoxin' was given to it, an acronym derived from the name of the second mentioned fungus. Further elucidation of the toxic factor demonstrated that the material could be separated chromatographically into four distinct spots.[9,10] All four components have been given the name 'aflatoxins' in order to identify their generic origin. Distinction of the four substances was made on the basis of their fluorescent colour with subscripts relating to the relative chromatographic mobility. Later on it became clear that the group of aflatoxins consists of at least 17 closely related compounds.

In the two decades following the outbreak of Turkey X Disease, a wealth of information about aflatoxins has been produced and many other mycotoxins have also been isolated and characterised. At present over 200

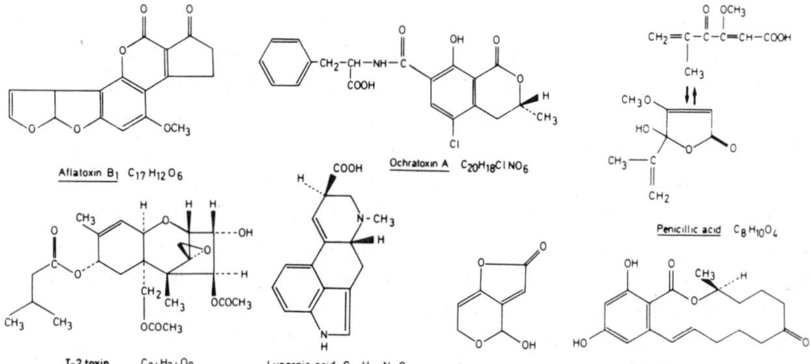

FIG. 1. Chemical structures of some mycotoxins.

different mycotoxins are known, showing a large variety of chemical structures. Examples of a few of these structures are shown in Fig. 1. The many different mycotoxins have demonstrated different biological effects in laboratory animals: acute toxic, mutagenic, carcinogenic, teratogenic, hallucinogenic, emetic and oestrogenic.

Humans may be exposed to mycotoxins through ingestion of toxin-contaminated food, inhalation or skin contact. The presence of mycotoxins in food may be the result of:

1. Direct fungal contamination of agricultural crops in the field, raw materials, manufactured products and final products. An example of field spoilage is the occurrence of aflatoxins in peanuts, whereas an example of spoilage of a semi-manufactured product is the occurrence of sterigmatocystin in cheese.
2. The contamination of animal products caused by contamination of the feedingstuff consumed by the animal. An example is the contamination of milk and dairy products with the 4-hydroxy derivative of aflatoxin B_1, called aflatoxin M_1, a metabolite formed through the dairy cow after ingestion of aflatoxin B_1 with the feedingstuff.[11]

Recent reviews about the occurrence of mycotoxins in various types of foodstuffs have appeared in the books *Environmental Health Criteria 11*[12] and *Mykotoxine in Lebensmitteln.*[13]

An example of human exposure through inhalation is the observation of Van Nieuwenhuize,[14] who reported the development of cancers in various organs of workers who had been inhaling small dust particles loaded with aflatoxins for several years in an oil-mill crushing peanuts and other oil seeds. A very dramatic example of exposure, by inhalation and skin contact, is the alleged use of mycotoxins in biological warfare.[15,16] The US Government has suspected that trichothecenes were being used in chemical attacks in South-East Asia. After so-called 'yellow rain' attacks, symptoms such as nausea, vomiting and diarrhoea were described by the victims, quite often civilians in Laos and Kampuchea.

Because of the widespread occurrence of the real or potential hazard of mycotoxins to human and animal health, many countries have enacted legal measures to control mycotoxin contamination of foodstuffs.[17] Because mycotoxins are natural contaminants which in most cases cannot be completely eliminated without outlawing the susceptible food or feed, public health officials are forced to make regulatory judgements. These judgements are particularly difficult when the mycotoxin, as is the case with aflatoxin, is shown to be a potent carcinogen in test animals, and the currently prevailing school of thought concerning carcinogens is that there is no level below which some effect cannot be anticipated.

For surveys, monitoring and enforcement programmes, methods of analysis are needed. The simplicity of the method will influence the amount of data that will be generated and the practicality of the ultimate control measures consequently taken. It may be clear that the availability of methods of analysis plays a key role in survey and research programmes. Nevertheless, efficient tackling of the mycotoxin problem requires a multidisciplinary approach, in which mycology, toxicology and chemistry each play a role of major importance. As it goes beyond the scope of this book which deals with food analysis, no further attention is paid to the mycological and toxicological aspects. It is not even possible to be complete as far as the determination of mycotoxins is concerned, since many hundreds of publications relative to the subject have appeared in recent years. However, in the following sections of this chapter the reader is introduced to the various aspects of mycotoxin analysis of foods to give an insight into analytical approaches, the problems that may arise, and the advantages and disadvantages of different types of methods of analysis. Special emphasis is laid on the determination of the aflatoxins, a group of highly carcinogenic mycotoxins, which have been intensively investigated.

2. SAMPLING AND SAMPLE PREPARATION

Sampling is an integral part of the analytical procedure. The object of the sampling procedure is to obtain a laboratory sample (test portion) representative of the lot from which it is drawn. Normally, the decision whether to accept or reject a lot is based on the evidence gained from analysis of the sample. When mycotoxins are homogeneously distributed throughout the lot to be inspected, sampling is made easy. A homogeneous distribution is encountered in the case of aflatoxin M_1 in milk and milk products because of the original fluid nature of these products. This situation is exceptional. Unfortunately, most mycotoxins are heterogeneously distributed and they may occur only in a fraction of the components of the batch to be inspected. Examples are the very uneven distribution of aflatoxin B_1 in a batch of peanuts[18] and some other particulate commodities, such as grain. There is a gradient distribution of sterigmatocystin in cheese, resulting from fungal contamination of the outer part of cheeses.[19] Because the distribution of aflatoxin B_1 in peanuts poses the greatest problem, it has been studied rather extensively, and this example will be used further throughout this section to demonstrate the difficulties in sampling as well as the approach to practical sampling procedures, despite the problems.

The total error made in a test procedure consists of three parts, namely the sampling error, the subsampling error (subsampling means that the original sample is comminuted, followed by sampling this ground fraction) and the analytical error. On the basis of a large number of analyses, Whitaker[20] was able to calculate the contribution of each error to the total error when a lot of peanuts contaminated with aflatoxin B_1 was sampled and analysed for aflatoxin B_1. As demonstrated in Fig. 2, the major error component is the sampling error, whereas the subsampling and (intra-laboratory) analysis error vary only slightly across all concentrations. (N.B. The inter-laboratory analytical error depends much more on the concentration.)

It is possible to draw a curve indicating the relationship of the probability of acceptance of a lot versus the aflatoxin concentration in the lot, given a certain tolerance. An example of such an operating characteristic (OC) curve is shown in Fig. 3.[20] From Fig. 3 it can be seen that the probability of accepting a lot approaches 1 when the concentration of aflatoxin approaches zero; and as the concentration becomes large, the probability of accepting approaches zero. Further, it is apparent from Fig. 3 that there

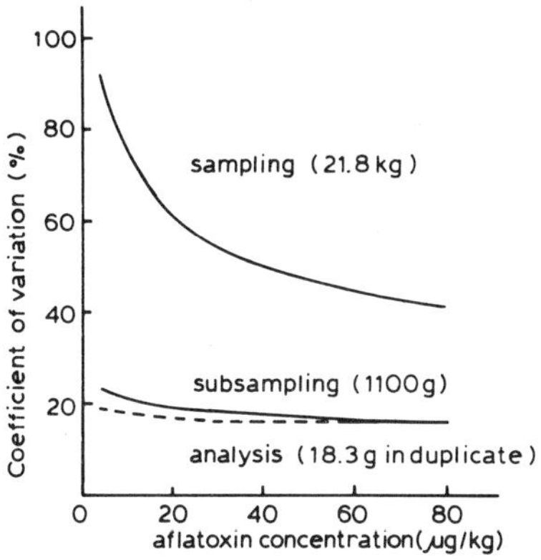

FIG. 2. Relative contributions of the errors of sampling, subsampling and analysis to the total error (after Whitaker[20]; courtesy IUPAC).

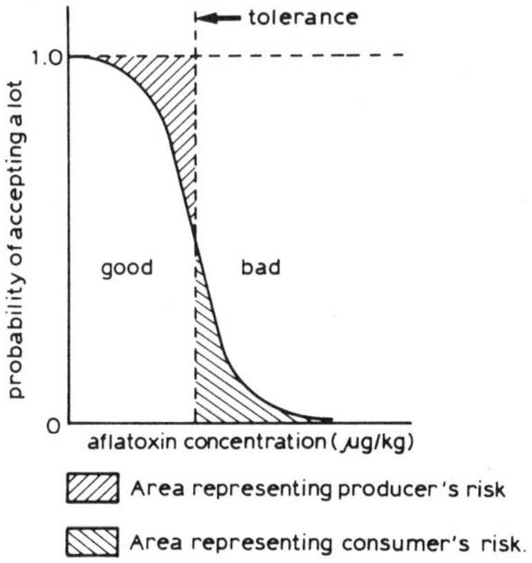

FIG. 3. Relation of the probability of acceptance of a lot versus the aflatoxin concentration (operating characteristic curve) (after Whitaker[20]; courtesy IUPAC).

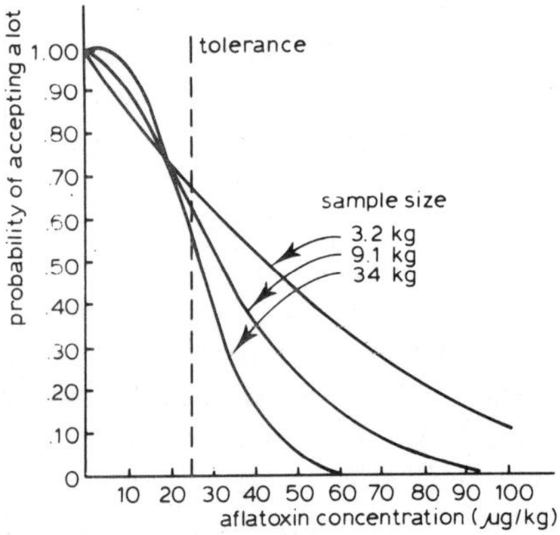

FIG. 4. Effect of sample size on the operating characteristic curve (after Dickens[21]; courtesy Institut für Toxikologie, Zürich).

are two risks, the producer's risk and the consumer's risk. The producer's risk is the risk that the lot will be falsely rejected, because the aflatoxin content measured in the test portion is higher than the tolerance, although the mean concentration in the lot is below the tolerance. The consumer's risk is the risk that a lot will be falsely accepted, because the aflatoxin content measured in the test portion is lower than the tolerance, although the mean concentration in the lot exceeds the tolerance. Increasing the sample size will lead to a reduction of both the consumer's risk and the producer's risk (Fig. 4).[21] The ideal OC curve is obtained when the whole lot is ground and analysed (Fig. 5).[21] Obviously the theoretically ideal situation has very impracticable consequences: nothing would be left to sell or to buy, at least not in its original form. The choice of the sample size depends on the risks that can be accepted and the costs one is willing to bear. Another possibility of influencing the producer's risk and the consumer's risk is changing the level of decision (this is a critical toxin level which has the following meaning: if the result of analysis of the test portion exceeds the limit of decision, the lot will be rejected). Lowering the level of decision reduces the consumer's risk; however, it leads to an increase in the producer's risk (Fig. 6).[21] To limit these risks, sampling plans have been developed in which two (or more) decision levels are used, an acceptance level and a rejection level. In such cases a lot is accepted if the outcome of the analysis of the test portion is lower than the acceptance level, rejected if it is higher than the rejection level, and re-analysed when the outcome is in between the two levels.

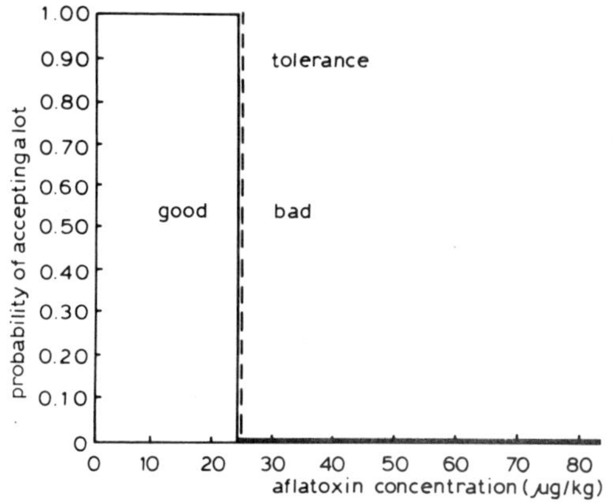

FIG. 5. Ideal operating characteristic curve: the whole lot is ground and analysed (after Dickens[21]; courtesy Institut für Toxikologie, Zürich).

An example of such a sampling plan is the PAC (Peanut Administrative Committee) sampling plan practised in the USA for sampling large lots of peanuts before they are shipped to the manufacturer.[22] This sampling procedure involves multiple sampling and assay from representative units of 22 kg of the lot and a tolerance for aflatoxin of 25 µg/kg. The PAC sampling plan is based on the experience that aflatoxin is statistically distributed in shelled peanuts according to the negative binomial distribution.[23] Another statistical model used to describe the distribution

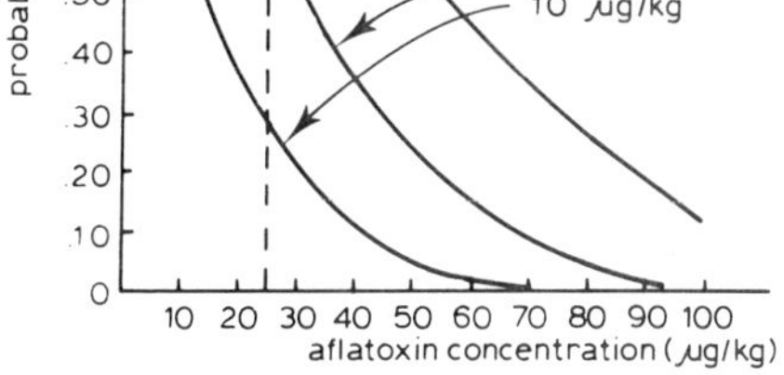

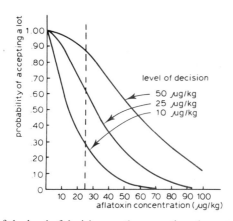

FIG. 6. Effect of the level of decision on the operating characteristic curve (after Dickens[21]; courtesy Institut für Toxikologie, Zürich).

of aflatoxin in peanuts is the compound Poisson-gamma distribution.[24] It is beyond the scope of this chapter to go into details of the difficult mathematical aspects of these models; those interested are advised to study the appropriate literature as referred to. Both the forementioned statistical models seem to fit quite well to the actual distribution of aflatoxin in peanuts when large samples are taken, and relatively high contamination levels are accepted as in the PAC testing programme. However, if the tolerance is only 1–5 μg/kg, as in several European countries,[17] and if decisions have to be made based on the analysis of small samples (as is the case in repressive control), there is a significant difference between the two distributions. Statisticians still disagree about which sampling procedure should then be used to obtain the best estimate of actual toxin content.

Although it is sometimes costly, there is no doubt that very large samples of many kilograms of peanuts must be taken to obtain a low risk of a wrong decision for both the consumer and the producer. These large sample sizes require the samples to be subsampled to make an adequate analysis possible. Because of the possible inhomogeneity, the whole sample must be ground and homogenised. For this purpose, special instruments and techniques have been developed, such as the Dickens–Satterwhite subsampling mill[25] and the Hobart vertical cutter–mixer.[26] Then, either the whole subsample is analysed or the size of the subsample is further reduced until a test portion, in size generally ranging from 20 to 100 g, is obtained. The compromise between solvent economy and a representative sample appears to have been set at 50 g. In the PAC testing programme for peanuts, the entire subsample (1100 g) is extracted with a mixture of 1650 ml of methanol, 1350 ml of water, 1000 ml of hexane and 22 g of sodium chloride. In addition to being costly, the solvents are an important energy resource and the used solvents are difficult to dispose of without environmental pollution. Therefore, Whitaker *et al.*[27] have proposed a water slurry method which consists of extracting aflatoxin with solvent from a 130 g sample of a slurry formed by blending 1100 g of comminuted peanut kernels, 1500 ml of water and 22 g of sodium chloride in a Waring Blender. It seems that the variance among analyses using the slurry method does not differ significantly from the variance among analyses using the official PAC procedure.

As well as the USA where the PAC sampling plan is practised, there were six other countries in 1981 which indicated having developed sampling plans for the control of mycotoxins (solely for aflatoxins).[17] It must be assumed that the design of these sampling plans would have been based on information about the following essential factors: a critical level (control, tolerance, guideline, etc., for aflatoxin); a definition of a good (acceptable) and a bad (rejectable) lot; and a statement of the acceptable or desired consumer's and producer's risks. In the absence of this information the selection of any sampling plan will be arbitrary.

3. ANALYTICAL TECHNIQUES

There are two approaches possible for the detection and determination of mycotoxins: biological and chemical. Biological methods may be useful in screening for known and unknown mycotoxins. As an example, they have played a role of importance in the period of the initial discovery of the aflatoxins.[28] However, if it is known which mycotoxin(s) should be looked for, chemical assays, if available, are to be preferred, because these generally are much more specific, more rapid, more reproducible, and possess lower limits of detection. Hence, chemical assays play a role of major importance in the determination of mycotoxins. Therefore, the bioassays are briefly discussed, whereas the chemical assays are described in more detail.

3.1. Bioassays

Generally, five categories of organisms are applied in bioassay systems: micro-organisms, aquatic animals, terrestrial animals, organ and tissue culture systems, and plants. An overview of the usefulness of these organisms for the detection of mycotoxins is given by Watson and Lindsay.[29]

Micro-organisms seem to be rather insensitive to mycotoxins. Burmeister and Hesseltine[30] surveyed over 300 species of micro-organisms for their sensitivity to aflatoxins and found only one strain of *Bacillus brevis* and two of *Bacillus megaterium* to be sensitive to aflatoxins. However, an assay method based upon the observed antibacterial action[31] had a high absolute limit of detection (1 μg) and was not sufficiently reproducible in collaborative studies to warrant further investigation. Other mycotoxins can inhibit the growth of micro-organisms in such a way that useful assay procedures may be developed.

Some aquatic animals, such as brine shrimp (*Artemia salina*) and certain fish species (trout, zebrafish, guppy), are used in bioassay systems to detect mycotoxins.[32] Generally the larvae of the fish or the brine shrimps are exposed to various concentrations of toxins dissolved in aquarium water and after a certain time the percentage kill is estimated for each dilution.[33]

The brine shrimp test is one of the simplest assays, but it is relatively insensitive to mycotoxins, except for some trichothecenes, sterigmatocystin and aflatoxins. A problem of the brine shrimp test may be the fact that many mycotoxins are only slightly soluble in water. The time required for the test is approximately one day and expertise is not required.

Among the terrestrial animals, ducklings and chick embryos seem to be the most sensitive to mycotoxins. In the original biological test developed during the outbreak of Turkey X Disease,[28] newly hatched ducklings were used as the test animal for determining the presence of aflatoxin isolated from suspect food, with bile duct hyperplasia as the specific, measured response. The lowest dose level of 0·4 μg/day administered for 5 days represents the minimum intake required to induce a detectable bile duct lesion. In the chick embryo assay a small amount of extract is introduced by means of a syringe at the side of the air cell. After incubation for 3–4 weeks the number of survivors is counted. The chick embryo test appears to be one of the most sensitive bioassay systems for mycotoxins. In addition, the test is reproducible and especially useful for aflatoxin B_1 assay, as typical lesions are observed in the embryo with subacute levels of aflatoxin B_1, less than 0·1 μg/egg. The chick embryo test has been studied collaboratively with success[34] and the method has been adopted as the official final action method by the Association of Official Analytical Chemists.[35] The long incubation period makes the chick embryo test one of the slower bioassays, which is the major disadvantage. When screening for trichothecenes, a group of mycotoxins exhibiting dermatitic activity, the laboratory animal skin test has proven to be useful.[36] Extracts of suspected commodities and cultures are applied on the shaved skin of a rabbit or a rat and the skin is inspected for some days. Responses such as erythema, oedema and necrosis indicate the presence of trichothecenes in the extract. The method is reliable and has an absolute limit of detection of *ca.* 0·01 μg/test.

The advantage of using cell cultures to detect mycotoxins is that low concentrations can give a perceptible response. Cultures of liver, kidney and muscle cells may serve as test material. This *in vitro* assay is carried out by adding extracts of suspect commodities or fungal cultures to the culture medium. Cultivation is continued for several days and the degree of cytotoxicity as well as cytogenetic effects and morphological changes are noted at certain time intervals.[37] Exposure of rat liver cells to concentrations of 0·1 μg aflatoxin B_1/ml medium leads to marked damage.[38] A limitation of the cell culture method is the fact that many media used for culturing fungi appear to be toxic and therefore they may not serve as a control.

The last category to be mentioned are the plants. The phytotoxicity assays are based on the ability of some mycotoxins to inhibit the growth and germination of seeds of higher plants. Schoental and White[39] found that watery suspensions of aflatoxins with a concentration of 25 μg/ml added to agar plates containing watercress seeds led to complete inhibition of seed germination, whereas chlorophyll deficiency in the seedlings ('albinism') was found at concentrations of 1–2·5 μg/ml. Burmeister and Hesseltine[40] reported that 0·5 μg of T-2 toxin inhibited the germination of pea seed by 50 % when the seeds were soaked overnight in the solution.

Bioassays may be useful when there is no chemical assay available, and have proved of primary use in screening for mycotoxins. However, their use in the surveillance of food and foodstuffs is of minor importance as they generally lack specificity, reproducibility and rapidity.

3.2. Chemical Assays

The limitations of bioassay techniques to detect and determine mycotoxins led chemists to develop more selective and reliable methods of analysis. Generally, all chemical analytical methods for the detection and determination of mycotoxins contain the basic steps as outlined in Fig. 7. The problems in sampling have already been discussed. Test portions that are normally used for analysis vary from 20 to 100 g.

3.2.1. Extraction

The first step in chemical analysis involves extraction of the test portion to isolate the component of interest. Generally, mycotoxins are extracted with

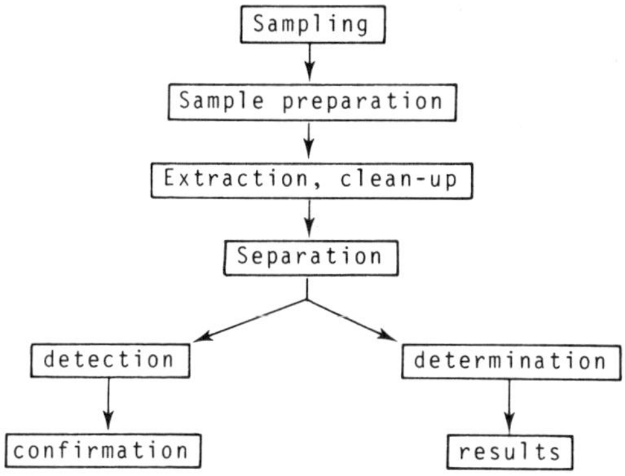

FIG. 7. Analytical procedure for mycotoxin determination.

organic solvents such as chloroform, dichloromethane, acetonitrile, ethylacetate, acetone and methanol. Contact between solvent and substrate is accomplished either for a short period (1–3 min) in a high-speed blender, or for a longer period (30 min) by shaking in a flask.

The choice of the solvent depends on the chemical properties of the toxin to be extracted as well as on the properties of the matrix. Often, mixtures of solvents, or solvents with small amounts of water and acids, are found to be most efficient. While the solubility of many mycotoxins in water is low, aqueous solvents may penetrate hydrophilic tissues, leading to a more efficient extraction by the non-aqueous solvents. One of the best-known and practised methods of analysis for aflatoxins, the CB (Contaminants Branch) method, employs a mixture of chloroform and water to extract aflatoxins.[41] Although the solubility of aflatoxins in water is low, the partitioning of aflatoxins from water to chloroform occurs rapidly.

3.2.2. Clean-up

Since mycotoxins are normally only present at very low levels, a strong concentration of the extract is necessary to make detection possible. The frequent presence of lipids and other interfering substances makes it necessary to clean-up the extract prior to concentration, by column clean-up and/or liquid–liquid extraction. Several column chromatographic clean-up steps are possible with materials such as silica gel, aluminium oxide, polyamide, Florisil® and Sephadex.® Silica gel is the most frequently used. Columns can be packed in the laboratory. However, prepacked columns are now commercially obtainable. The advantages of such prepacked columns, e.g. Sep-pak® and Baker®, are obvious. Variations in preparation of columns between analysts are eliminated and the time needed to prepare the columns is saved. On the other hand, prepacked columns do not offer the possibility of easily introducing slight variations in the column composition (for instance, adjustment of the water content).

The sample extract is usually added to the column in an appropriate solvent, after which the column is washed with one or more solvents in which the toxins are insoluble or less soluble than the impurities. Then the solvent composition is changed in such a way that the toxins are eluted from the column. The eluate is collected and concentrated. The concentration step is usually accomplished by evaporating the solvent in a rotary evaporator under reduced pressure, or by using a steam bath, while keeping the sample under a stream of nitrogen. The residue is redissolved in a small volume of solvent, quantitatively transferred to a small vial, brought to a specified volume, and this final extract is saved for detection and quantitation of toxin present.

Liquid–liquid extraction may also be carried out in separating funnels, for instance pentane against methanol–water. Since most mycotoxins are not lipophilic, fats can be removed in this way without toxin. The above-mentioned clean-up techniques are in fact separation procedures in which groups of substances with certain physico-chemical properties can be separated from one another. In this way the greater part of the co-extracted material can be removed. The choice of clean-up procedure depends on the method used for detection and determination, the required limit of detection, the speed of analysis and the recovery.

3.2.3. Ultimate Separation, Detection and Determination

Despite extraction and clean-up, the final extract may contain large amounts of other co-extracted substances possibly interfering with mycotoxin determination. Several possibilities exist to separate the mycotoxins from the matrix. Chromatographic procedures, which are physical separation techniques, are most often applied and they are used in combination with visual or instrumental determination of the mycotoxin(s) of interest. Immunoassays, which are biochemical separation techniques used in combination with instrumental determination, are still in an early stage of development for mycotoxin research. Nevertheless, the latter techniques are promising and it is to be expected that these will become of increasing importance.

(*a*) *Chromatographic Procedures*

In the determination of mycotoxins, adsorption chromatography and partition chromatography are the most important types of chromatography. For most of the time the phenomenon on which the separation is based is a combination of both types which can be subdivided into:

(i) open-column chromatography;
(ii) thin-layer chromatography;
(iii) high-performance liquid chromatography;
(iv) gas–liquid chromatography.

(i) *Open-column chromatography.* Open-column chromatography has already been mentioned as a technique often used in clean-up procedures. A special design—the glass mini-column with an internal diameter of *ca.* 5 mm—can be used for the detection of some mycotoxins in certain commodities. The first mini-column, introduced by Holaday[42] for the detection of aflatoxins in peanuts, contains a 45 mm silica gel layer held in place by glass wool at both ends. The column is dipped in the extract to be analysed and drawn up by capillary force. Depending on their physico-chemical characteristics the extract components move to a greater or lesser extent with the migrating solvent and separated bands appear on the column. After 10–15 min the column is examined under UV light for the characteristic blue or bluish-green colour that aflatoxins emit when excited by longwave (365 nm) UV light. The method of Holaday[42] has been refined and improved, especially as far as the column packing is concerned. Packings with successive zones of adsorbents such as alumina, silica gel and Florisil® with calcium sulphate drier at both ends and held in place with glass wool (Fig. 8) have proved to be effective and very useful in the analysis of commodities for aflatoxins.[43] Contrary to the ascending chromatography in the original method of Holaday,[42] descending chromatography

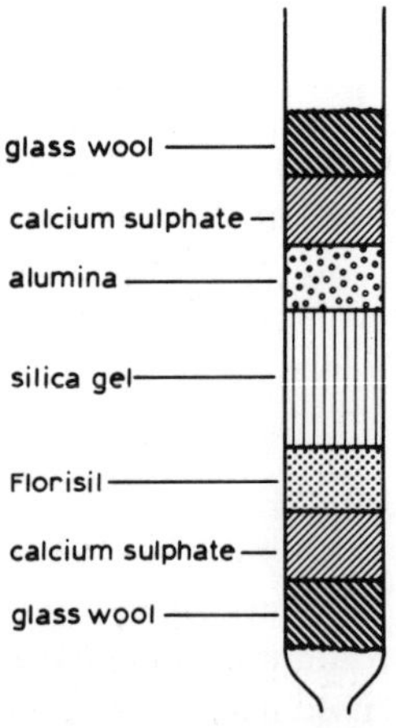

FIG. 8. Packing of a mini-column according to Romer.[43]

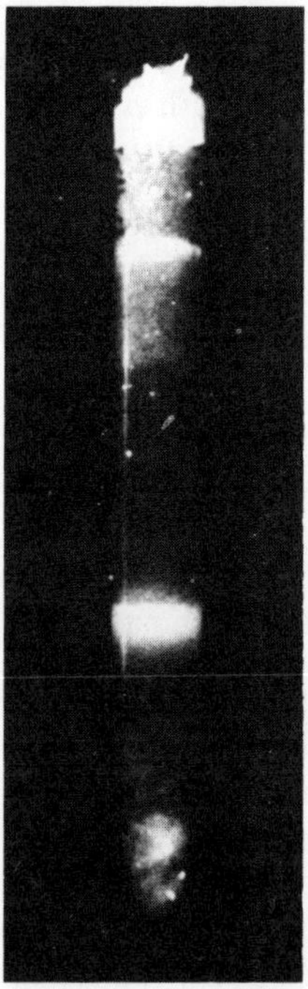

FIG. 9. Adsorption of aflatoxin B_1 to the Florisil® layer of a mini-column.

with a mixture of chloroform and acetone is applied in the method of Romer,[43] trapping the aflatoxins as a tight band at the top of the Florisil® layer, where they can be detected by their blue fluorescence under UV light (Fig. 9). By comparing a sample column with a column containing a known amount of aflatoxin, it is possible to judge whether the sample contains more or less aflatoxin than the standard.

The method of Romer[43] was subjected to a successful collaborative study[44] and has been adopted by the AOAC as an official first action method for the detection of aflatoxins in almonds, white and yellow maize, peanut and cotton seed meals, peanuts, peanut butter, pistachio nuts and mixed feeds.[35] Contrary to thin-layer chromatographic techniques (see Fig. 10), the mini-column method of Romer[43] does not distinguish between the different aflatoxins. As well as for the aflatoxins, similar mini-column procedures have been developed for some other mycotoxins that fluoresce when irradiated with UV light, such as ochratoxin A (see Fig. 1) in a wide range of products[45] and zearalenone (see Fig. 1) in maize, wheat and sorghum.[46] The limits of detection achieved vary from *ca.* 5–15 μg/kg for the aflatoxins to *ca.* 20 μg/kg for ochratoxin A and *ca.* 50 μg/kg for zearalenone.

Mini-column methods are 'go/no go' methods, which require only little time and expertise and no sophisticated equipment. This makes them particularly useful for field screening tests by scientists and technicians in developing countries. Therefore, the Organisation for Economic Cooperation and Development (OECD) has published some selected mini-column procedures for aflatoxins in a *Handbook on Rapid Detection of Mycotoxins*.[47] Although having some advantages, mini-column methods have certain limitations. They are at best semi-quantitative and generally have a higher limit of detection and less sensitivity, separation power and selectivity than is obtained by using thin-layer chromatographic and high-performance liquid chromatographic procedures.

(ii) *Thin-layer chromatography*. In the first years of mycotoxin research, thin-layer chromatography (TLC) became a very common and popular technique for separating extract components and nowadays there are still numerous applications. Initially separations were carried out in one dimension using a single developing solvent. Later, two-dimensional TLC was introduced to mycotoxin research;[48] it is a powerful separation technique in which a second development is carried out in a direction at right angles to the first one, using a different developing solvent. This provides a much better separation than one-dimensional TLC and is required especially in those cases where low levels of detection are necessary, e.g. aflatoxin M_1 in milk, and if extracts contain many interfering substances, e.g. feedingstuffs, roasted peanuts.

In thin-layer chromatography a wide range of adsorbents can be used. For mycotoxin research silica gel TLC plates are most often used as this type of adsorbent generally offers the best possibility of separating the toxin of interest from matrix components. Both precoated and self-coated plates can be used. Self-coated plates allow a free selection of adsorbent and a free selection of additives. Calcium sulphate can be added as a binder of the silica gel to the glass plate. EDTA has been used by Stubblefield[49] as a complexing agent for contaminants in the silica gel to prevent streaking of citrinin spots. Precoated plates, on the other hand, are ready to use, they generally possess a more uniform and rigid layer and do permit a certain choice of support, e.g. glass, plastic or aluminium. The characteristics of precoated as well as self-coated plates may differ from brand to brand and sometimes even from batch to batch, leading to different separation behaviour, as can be seen in Fig. 10, where a mixture of aflatoxins B_1, B_2, G_1 and G_2 has been separated on three different types of plates, after which the plates are viewed under UV light. (These four main aflatoxins have been

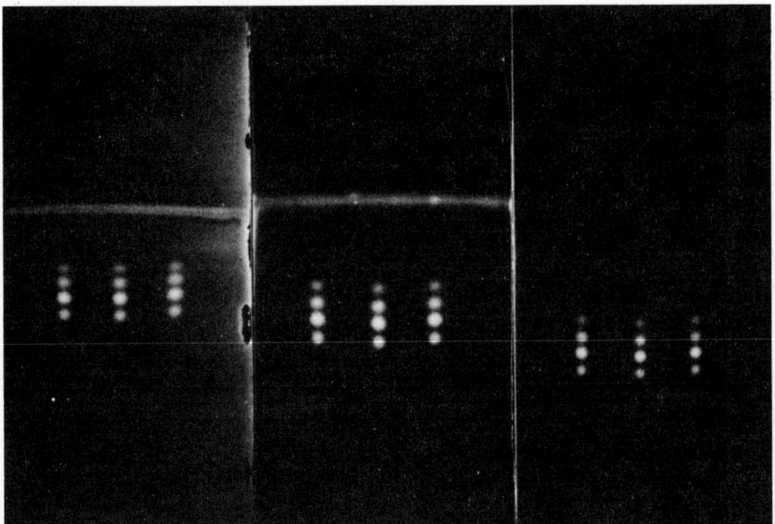

FIG. 10. Separation of a mixture of aflatoxin standard on several types of SiO_2 TLC plates.

named according to their colour of fluorescence—blue or green—and their R_f value.)

Thin-layer plates can be used in different formats. Most separation problems may be resolved using a square 20 × 20 cm plate; however, the use of 10 × 10 cm plates and even 7 × 7 cm self-cut plates will often lead to good results as well. For two-dimensional separation procedures especially, the use of the smaller sizes saves much time. Examples of analytical procedures in which two-dimensional separations are carried out on small TLC plates are the multi-mycotoxin method of Patterson and Roberts,[50] the method of Van Egmond *et al.*[51] for the determination of sterigmatocystin in cheese, the method of Stubblefield and Shotwell[52] for the determination of aflatoxins in animal tissue, and the method of Paulsch *et al.*[53] for the determination of ochratoxin A in pig kidneys. The smaller plate sizes additionally offer the possibility of developing the plates not only in specially designed developing chambers, but also in simple beakers or in series of ten at a time in a so-called multi-plate rack (Fig. 11), thus significantly reducing the overall time required for a series of TLC runs.

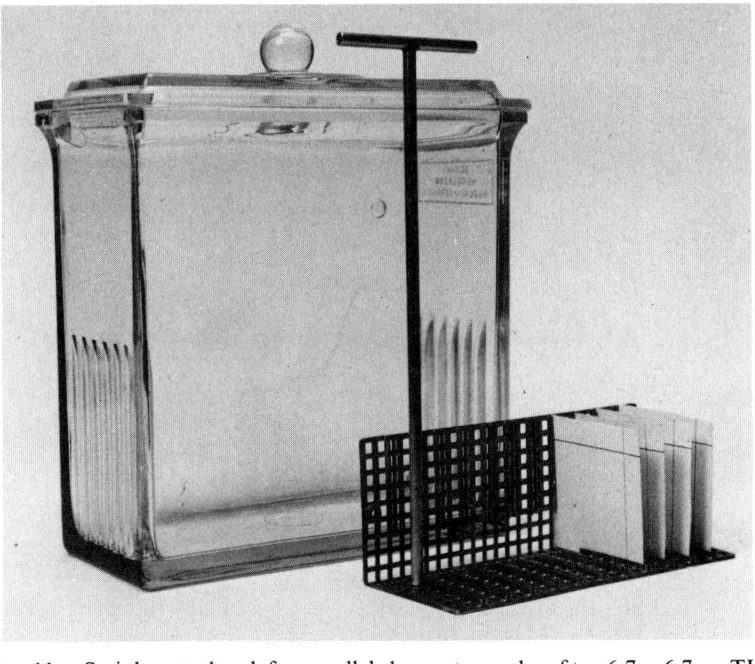

FIG. 11. Stainless steel rack for parallel chromatography of ten 6·7 × 6·7 cm TLC plates.

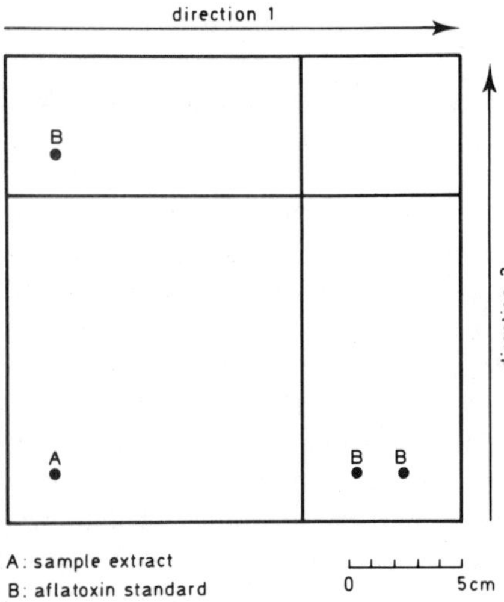

FIG. 12. Scheme of the spotting pattern for two-dimensional TLC (densitometric quantification).

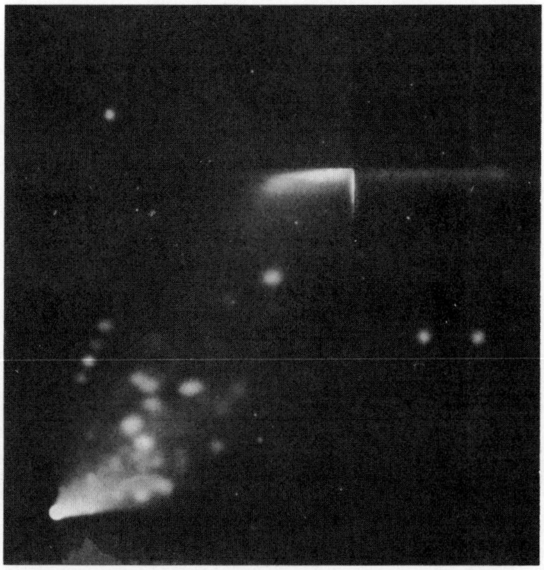

FIG. 13. Separation of an extract of peanut butter submitted to two-dimensional TLC, using the densitometric spotting pattern.

In the TLC determination of mycotoxins, generally 5–25 μl of extract is applied to the plate. Depending on the desired accuracy and precision, different types of applicators are used. For screening purposes the disposable qualitative capillary pipettes will suffice. For quantitative work disposable quantitative capillary pipettes or precision syringes, which are more accurate and precise, are used. Moreover, the latter allow the intermittent application of larger volumes under inert atmosphere by using them in combination with a repeating dispenser incorporated in a spotting device. The spotting of sample and standard(s) is normally carried out according to a spotting pattern, prescribed as a part of the whole analytical procedure. Different spotting patterns apply to one-dimensional TLC or, in the case of 'dirty' extracts, two-dimensional TLC. In two-dimensional TLC the sample extract is spotted at a corner of the TLC plate and two developments are carried out successively parallel to the two sides of the plate using two different developing solvents. The two solvents must be compatible and independent, i.e. there should be little correlation between the retention patterns in both systems, otherwise the spots tend to agglomerate along the bisector of the plate. An extensive study of the theoretical aspects of two-dimensional TLC has been made by Guiochon *et al.*[54]

An example of the use of two-dimensional TLC is the procedure used in the official EEC method for the determination of aflatoxin B_1 in feedingstuffs[55] (Fig. 12). An aliquot of extract is spotted at A and known amounts of aflatoxin B_1 standard are spotted at B. The plate is then developed in the first direction with a mixture of diethyl ether, methanol and water (94:4·5:1·5) and, after drying, the plate is turned 90° and developed in the second direction with a mixture of chloroform and acetone (9:1). Detection and quantification are carried out under longwave UV light (365 nm). In Fig. 13 the result of a two-dimensional TLC separation of an extract of peanut butter contaminated with aflatoxin B_1 is shown. With the help of the co-developed B_1 standards, the well-separated B_1 spot from the sample can be located. By means of a densitometer the intensities of fluorescence of the B_1 spot from sample and standard can be compared and thus the B_1 concentration in the initial sample can be calculated.

If a densitometer is not available, an anti-diagonal spotting pattern may be used as originally developed by Beljaars *et al.*[56] for the determination of aflatoxin B_1 in peanuts (Fig. 14). An aliquot of sample extract is spotted at A and different amounts of B_1 standard are spotted at the points B. The plate is developed two-dimensionally, and detection and quantification are again carried out under UV light (Fig. 15). With the help

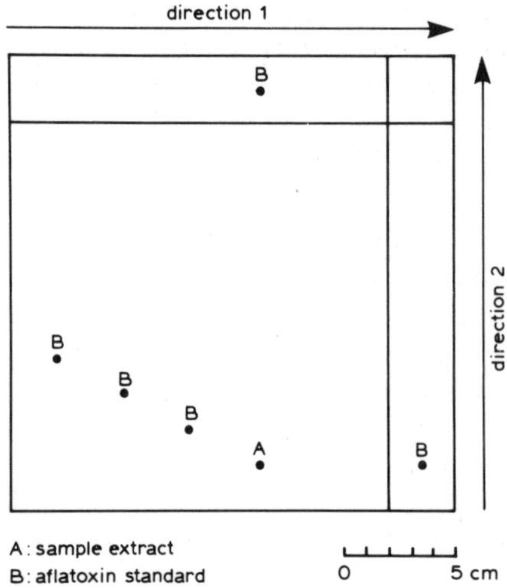

FIG. 14. Scheme of the anti-diagonal spotting pattern for two-dimensional TLC (visual quantification).

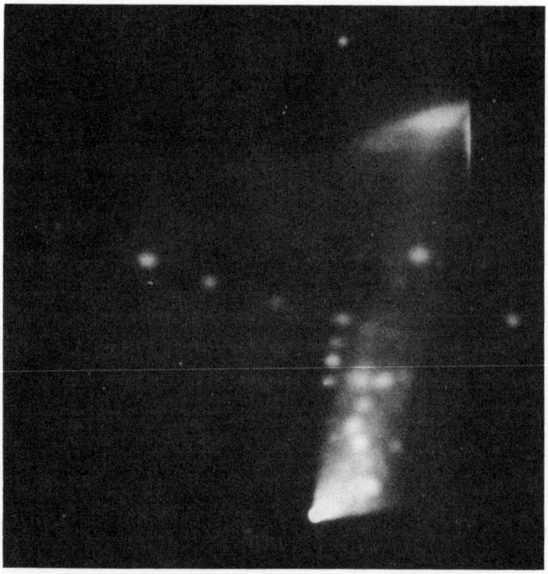

FIG. 15. Separation of an extract of peanut butter submitted to two-dimensional TLC, using the anti-diagonal spotting pattern.

of the row of two-dimensionally developed B_1 standards, B_1 from the extract can be located and its concentration estimated by comparing its intensity of fluorescence with that of the different B_1 standards. As all the B_1 spots are in a line and rather close to each other, such an estimation is easier than visual comparison with standard spots developed in the side lanes, as is the case in the densitometric spotting pattern. However, the technique is applicable only if the standard spots appear on a 'free' part of the plate after two-dimensional development. In Fig. 15 the result of a two-dimensional TLC separation of an extract of peanut butter is shown, using the anti-diagonal spotting pattern. Two-dimensional TLC procedures have also been developed for the determination of other mycotoxins such as zearalenone,[57] sterigmatocystin,[51] ochratoxin A[53] and in a multi-mycotoxin method.[50]

The fortunate characteristic that aflatoxins emit the energy of absorbed longwave UV light as fluorescent light enables the analyst to detect these compounds at low levels. Unfortunately, not all mycotoxins can be detected by such a simple method. Many do not fluoresce under UV light, some show UV or visible light absorption, while others do not. If the latter is the case, the mycotoxin can sometimes be made visible by spraying a reagent on the plate or by exposing the plate to reagent vapour. An example of such a derivatisation is the spraying technique used for the visualisation of sterigmatocystin, a toxin sometimes occurring in grains[58] and in cheese.[59] Stack and Rodricks[60] have found that spraying with an $AlCl_3$ solution leads to an Al complex with the keto- and hydroxyl groups of the sterigmatocystin molecule (Fig. 16), resulting in an enhancement of the fluorescence intensity of *ca.* 100 times. In addition, the colour of fluorescence changes from brick-red to yellow.

In spite of all the clean-up techniques used, there are still substances which behave in the same manner during TLC separation as the mycotoxin being determined. In order to minimise the risk of false-positives, the

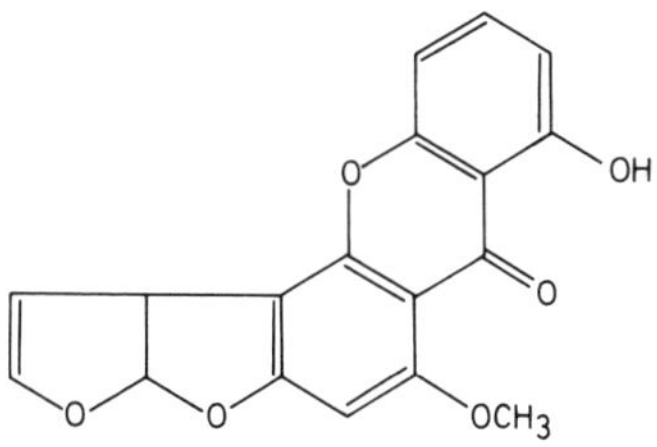

FIG. 16. Chemical structure of sterigmatocystin, $C_{18}H_{12}O_6$.

identity of the mycotoxin in positive samples has to be confirmed. The most reliable method for this purpose is high-resolution mass spectrometry (HRMS). HRMS in combination with TLC, however, is rather time-consuming and not every laboratory is equipped with this sophisticated type of apparatus. Therefore more simple techniques have to be applied. Probably the simplest ways of confirming the presence of mycotoxins are the use of additional solvent systems or the application of supplementary chromatography, by repeating the TLC procedure with an internal standard superimposed on the extract spot before developing the plate. After completion of TLC this superimposed standard and the 'presumed' toxin spot from the sample must coincide. Another possibility is to spray the developed TLC plate with a reagent, so that the colour (of fluorescence) of the mycotoxin spot changes. An example of the latter possibility is the spraying test with a dilute solution of sulphuric acid,[61] which leads to a change in the colour of fluorescence of aflatoxin spots from blue to yellow. Although the tests described above, if negative, would rule out the presence of the mycotoxin concerned, they do not provide positive confirmatory evidence.

Positive identification can be obtained by formation of specific derivatives with altered chromatographic properties. Both mycotoxin standard and suspected sample are submitted to the same derivatisation reaction. Consequently, in positive samples a derivative from the mycotoxin should appear, identical to the derivative from the mycotoxin standard. Confirmatory reactions may be carried out in test tubes or, preferably, directly on a TLC plate, thus using the separation power of TLC. The formation of chemical derivatives of mycotoxins used for confirmation purposes was first described for aflatoxin B_1 by Andrellos and Reid.[62] Mixing of aflatoxin B_1 with glacial acetic acid–thionyl chloride, formic acid–thionyl chloride or trifluoroacetic acid leads to fluorescent derivatives by reaction with the double bond of the terminal furan ring of aflatoxin B_1. Pohland *et al.*[63] investigated the acid-catalysed addition of water to the double bond of the terminal furan ring of aflatoxin B_1 and found that 2-hydroxydihydro-aflatoxin B_1 (B_{2a}) was formed, a blue fluorescent derivative with a specific lower R_f value on silica gel TLC plates than aflatoxin B_1 itself. Based on these findings, confirmation procedures for aflatoxin B_1 in extracts have been developed.

Initially[64] the reaction was carried out in a test tube with the sample extract as well as with aflatoxin B_1 standard solution, after which TLC was carried out to compare the R_f values of the respective derivatives. After it was established that the reaction could be carried out on the TLC plate a well, *in situ* derivatisation procedures were developed which are more practical. In the method of Przybylski,[65] trifluoroacetic acid is superimposed on the extract spot before development. After reaction the plate is developed and examined under UV light for the presence of the blue fluorescent spot of aflatoxin B_{2a}, which can be recognised with the help of an aflatoxin B_1 standard, spotted on the same plate, and which has undergone the same procedure. In the case of very 'dirty' extracts (e.g. from feedingstuffs) it may be difficult to notice the aflatoxin B_{2a} spot owing to background fluorescence. In such circumstances the method of Verhülsdonk *et al.*[66] is very useful. In this procedure a so-called separation–reaction–separation procedure is carried out (Fig. 17). Hydrochloric acid is sprayed after the first separation run, and the reaction takes place. Then a second separation is carried out in the second direction, under identical conditions, after which the isolated blue fluorescent spot of aflatoxin B_{2a} is visible, which can be recognised with the help of a B_1 standard, spotted on the same plate, which has undergone the same procedure. Other (unreacted) components lie on a diagonal line, bisecting the plate, as the separation was carried out in both directions under exactly identical conditions. In Fig. 18 the result of such a confirmatory test applied to rabbit feedingstuff contaminated with aflatoxins B_1 and G_1 is shown. The methods of Przybylski[65] as well as of Verhülsdonk *et al.*[66] have been adopted as confirmation tests in the official EEC method for the determination of aflatoxin in feedingstuffs.[55]

The techniques described for confirmation of aflatoxin B_1 are also applicable to aflatoxin G_1, but not to aflatoxins B_2 and G_2 of which the

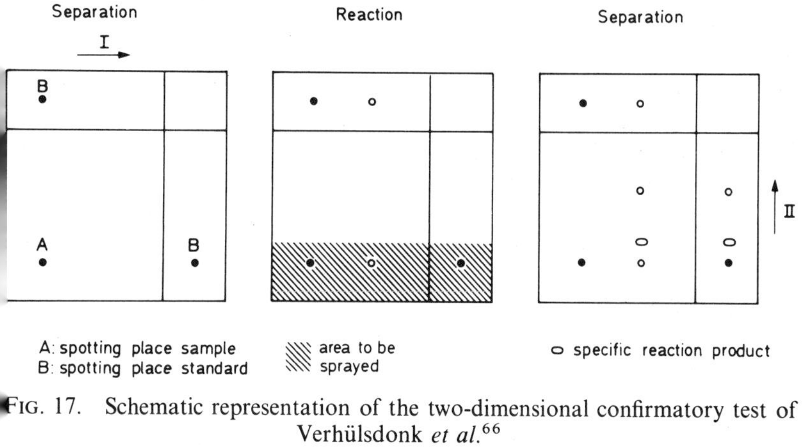

FIG. 17. Schematic representation of the two-dimensional confirmatory test of Verhülsdonk *et al.*[66]

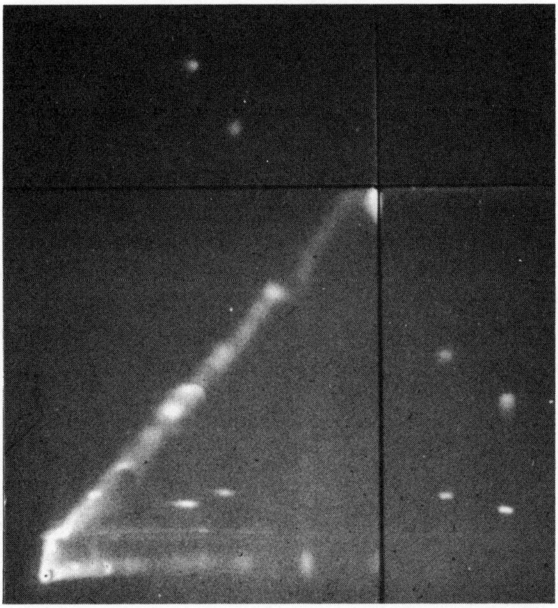

FIG. 18. Result of the confirmatory test of Verhülsdonk *et al.*,[66] applied to rabbit feedingstuff contaminated with aflatoxins B_1 and G_1.

terminal furan ring is saturated. In addition, an *in situ* confirmatory test has been developed for aflatoxin M_1 by Trucksess.[67] In this procedure a reaction is carried out between trifluoroacetic acid and aflatoxin M_1 on the origin spot of a TLC plate similar to the confirmation procedure for aflatoxin B_1 of Przybylski.[65] Contrary to the identity of the aflatoxin B_1 reaction product, the aflatoxin M_1 reaction product formed on the TLC plate has been shown not to be the hemiacetal, even though at the time of writing the product has not yet been identified. The method of Trucksess[67] has been modified by Van Egmond *et al.*[68,69] in order to make confirmation of identity possible at the very low aflatoxin M_1 concentrations more commonly found in milk in some European countries.

In situ derivatisation procedures on TLC plates followed by TLC of the reaction product(s) to establish the identity of mycotoxins other than aflatoxins are rather scarce. Van Egmond *et al.*[51] described a test for the confirmation of identity of sterigmatocystin (see Fig. 16) in a cheese extract. In the test, which is based on the principles of the separation–reaction–separation procedure (see Fig. 17), a mixture of trifluoroacetic acid and benzene is sprayed on the TLC plate and after the reaction and a second development the reaction product is visualised with $AlCl_3$ spray reagent.[60] Paulsch *et al.*[53] developed a confirmatory test for ochratoxin A (see Fig. 1) in pig kidneys, based on the findings of Kleinau.[70] Ochratoxin A spots separated after two-dimensional TLC are esterified on the TLC plate with methanol–H_2SO_4, after which the plate is developed for the third time. The methyl ester of ochratoxin A is visible under longwave UV light as a fluorescent spot with an R_f value higher than that of ochratoxin A. Like ochratoxin A, the methyl ester undergoes the same change in colour of fluorescence from green to blue when the pH of the plate is changed from acid to alkaline by exposing the plate to the vapour of ammonia, and this phenomenon can be considered as an additional confirmation of identity.

The use of thin-layer chromatography as a technique to separate mycotoxins from matrix components has decreased in recent years in favour of high-performance liquid chromatography and to a lesser extent of gas–liquid chromatography (especially for the determination of trichothecenes). Further analytical developments may be expected in the near future from immunoassays. Nevertheless, thin-layer chromatography is a reliable, relatively simple and still the most frequently used technique for the determination of mycotoxins, especially its two-dimensional application which offers a good resolution, and consequently low limits of detection. Special advantages of thin-layer chromatography are the possibility of carrying out *in situ* derivatisation procedures to confirm the presence of mycotoxins, the ability to store plates for later interpretation, and the fact that the analyst has a certain 'contact' with the results of the separation, because the human eye itself can act as a detector. A disadvantage with regard to the other techniques mentioned is the inability to automate the procedure easily. In summary, thin-layer chromatography is a major separation technique in mycotoxin research, and is particularly recommended to those who are inexperienced in the analysis of food and feed for mycotoxins, and who cannot afford to purchase sophisticated analytical instrumentation.

(iii) *High-performance liquid chromatography.* High-performance liquid chromatography (HPLC), or high-pressure liquid chromatography as it was initially called, became available for the analysis of foodstuffs in the early 1970s and probably the first published application for mycotoxin research dates from 1973.[71] After a somewhat hesitant start, the technique became of rapidly growing importance in the determination of mycotoxins, particularly when several types of column packings and (fluorescence) detectors became available. The introduction of autosamplers and computerised data retrieval systems made HPLC very useful in principle for large-scale analyses.

In the first applications which concerned aflatoxin assays, SiO_2 columns were used in combination with chloroform- or dichloromethane-containing mobile phases, and detection was by means of UV detectors ($\lambda = 254$ or 365 nm).[71-73] UV detection, however, is not very selective, whereas the obtained limits of detection for aflatoxins are relatively high (in the order of *ca.* 1–2 ng of each aflatoxin[72] or expressed as a relative detection limit, *ca.* 50 μg aflatoxin B_1/kg for groundnuts.[73] Therefore the use of UV detectors for aflatoxin assay was largely discontinued when the more selective fluorescence detectors became available. Initially the fluorescence detectors had limitations in detectability as well, because aflatoxins B_1 and B_2 do not exhibit strong fluorescence in normal-phase solvents and B_1 and G_1 do not exhibit strong fluorescence in reverse-phase solvents. Consequently the limits of detection of the aflatoxins mentioned could not compete with those obtained in thin-layer chromatography. Several efforts have been undertaken to improve the intensity of fluorescence of the aflatoxins, which resulted in four main techniques:

1. In the procedures of Takahashi[74] and Haghighi *et al.*,[75] the sample and standard solutions are treated with acid (i.e. trifluoroacetic acid) to convert aflatoxins B_1 and G_1 to the respective hemiacetals, B_{2a} and G_{2a}. The hemiacetals fluoresce as strongly as B_2 and G_2 in reverse-phase solvents. A disadvantage of the technique is that it involves an extra step needed for the chemical conversion. In addition, it may be questionable whether conversion always occurs quantitatively.

2. Panalaks and Scott[76] and Zimmerli[77] introduced the use of silica gel-packed flow cells for fluorimetric detection of aflatoxins in normal-phase solvents. In the adsorbed state, the aflatoxins B_1 and B_2 fluoresce much more intensively than they do in solution. The limits of detection thus obtained are of the same order of magnitude as those based upon TLC with fluorescence detection. The life expectancy of the packed flow cell may vary depending on the number and state of the samples that are injected. In practice, contamination of the flow cell occurs, caused by deposits from dirty extracts accumulating on the silica gel over a period of time, thus necessitating frequent changes. The latter is a restriction of the practical use of a packed flow cell, especially when the detector has no easily accessible flow cell. Panalaks and Scott[76] indicated that it was not convenient to regenerate the flow cell by pumping through a more polar solvent because this caused a change in the transparency of the silica gel.

3. Manabe *et al.*[78] introduced a new mobile phase which prevented the usual quenching of aflatoxins B_1 and B_2. A solvent consisting of a mixture of toluene, ethyl acetate, formic acid and methanol would lead to a minimum detectable amount of aflatoxins B_1 of *ca.* 0·3 ng, whereas application of the method to food and feed revealed that levels of 10–20 μg/kg of the four aflatoxins B_1, B_2, G_1 and G_2 were still detectable. Compared to the other techniques used to enhance the fluorescence of aflatoxins B_1 and B_2, the limits of detection claimed by Manabe *et al.*[78] are not too impressive; however, the technique is the easiest to apply. A disadvantage is the possible decomposition of aflatoxins B_1 and G_1 on the column due to the presence of acid in the mobile phase. Care should be taken not to use too high amounts of acid and to check, using standards, whether any decomposition occurs under the optimal separation conditions.

4. Treatment of aflatoxins B_1 and G_1 with iodine to produce more intensely fluorescing derivatives has been reported by Davis and Diener.[79] The finding was made applicable to reverse-phase HPLC by the same authors[80] and the application has been further refined by Thorpe *et al.*,[81] who described the post-column derivatisation to the detection of aflatoxins by reverse-phase HPLC. The iodine addition enhances the fluorescence of aflatoxins B_1 and G_1 approximately 50-fold without affecting the fluorescence of aflatoxins B_2 and G_2. The procedure has proved to be successful for the analysis of samples of corn and peanut butter, contaminated at levels ranging from 0·5 to 2·0 μg/kg. In addition to the fluorescence enhancement of aflatoxins B_1 and G_1, the advantages of the procedure of Thorpe *et al.*[81] include the need to derivatise only the portion of the sample injected into the liquid chromatograph with and without post-column reagent addition, which confirms the presence or absence of aflatoxins B_1 and G_1. At the time of writing, no specific disadvantages of the iodine derivatisation method have yet been mentioned.

Next to the use of conventional fluorescence detectors with or without extra provisions, the use of laser-induced fluorescence has been described for aflatoxin detection.[82] With this very sensitive technique, as little as 750 femtograms (750×10^{-9} μg) of aflatoxin B_1 could be detected in model experiments. When applied to a maize extract, however, a limit of detection of 2 μg/kg was obtained in practice.

HPLC methods have also become available for the analysis of milk and milk products for aflatoxin M_1.[83–85] Most of these methods use reverse-phase HPLC, which does not lead to problems in detectability, as aflatoxin M_1 fluoresces much more intensively in reverse-phase solvents than aflatoxin B_1, so that no special provisions or derivatisations are necessary. The limits of detection obtained are comparable with those obtained in (two-dimensional) thin-layer chromatography and some of these methods are widely used already in surveillance and monitoring programmes. An example of an HPLC separation of an extract of milk prepared according to the method of Tuinstra *et al.*[85] is shown in Fig. 19. As illustrated, detection at a level of contamination of 0·01 μg/kg is possible, although quantification at this level is a rather precarious matter.

Simultaneously with the development of new column types and refined HPLC equipment, several commercial firms have brought ready-packed disposable columns on to the market. These new columns have already been mentioned on p. 110. Some of the new column types such as the silica and C_{18} Sep-pak® cartridges have been incorporated in HPLC procedures for aflatoxins in soya products,[86] in maize and dairy feeds,[87] and for aflatoxin M_1 in milk.[84,88] In this way the speed of HPLC assays could be further enhanced, which indeed is not a specific advantage, as these types of columns may be used for the clean-up in TLC assays as well.

In the ten years following the first published application of HPLC analysis in mycotoxin research,[71] approximately 80 HPLC methods for

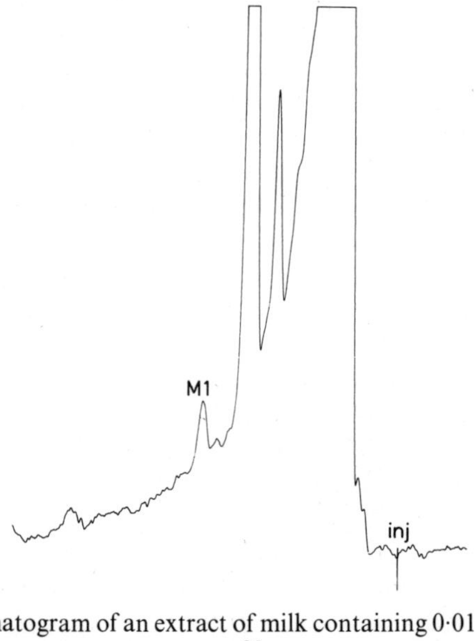

FIG. 19. Chromatogram of an extract of milk containing 0·01 μg M_1/kg prepared according to the method of Tuinstra *et al.*[85] (Courtesy Springer-Verlag, Heidelberg.)

aflatoxins have been published in those scientific journals that are covered by *Chemical Abstracts*. It is inappropriate to review them all here, especially as the number of publications dealing with the subject is still growing, so just a few examples of the various techniques and approaches are given. The same argument is valid for the other mycotoxins for which HPLC procedures have been developed, although this group is still less extensive. Without intentionally ignoring equally interesting methods, some procedures are worth mentioning.

Hunt *et al.*[89,90] described an HPLC method to detect and quantify ochratoxin A in pig kidney down to below 1 μg/kg. In the method an enzymatic digestion is carried out in order to liberate the toxin from the matrix as efficiently as possible. After extraction and clean-up the final extract is separated using a reverse-phase column. To confirm the identity of ochratoxin A in the tissue, two methods have been devised. In the first, a stream of ammonia is introduced into the eluant from the HPLC column. This produces an enhanced response relative to a separate injection made when the ammonia stream is switched off. In the second confirmatory method the methyl ester is prepared and subjected to chromatography under the same conditions. This produces an increase in retention time by comparison with the unesterified ochratoxin peak. Surveys of pig tissue have shown that artefacts are present which could be falsely recorded as ochratoxin positives without the use of the two confirmation techniques described above.[91]

A new development in the determination of trichothecenes (a group of 37 structurally related mycotoxins) (see Fig. 20) has been described by Cooper *et al.*[92] in which trichothecenes can be detected using chemical derivatisation and subsequent HPLC/UV detector analysis. The author claims a limit of detection of 1 μg/kg or lower with good recoveries from spiked samples. The analysis for trichothecenes has been a difficult task, although a breakthrough has been achieved through the recent application of capillary gas chromatography. A reliable HPLC procedure capable of

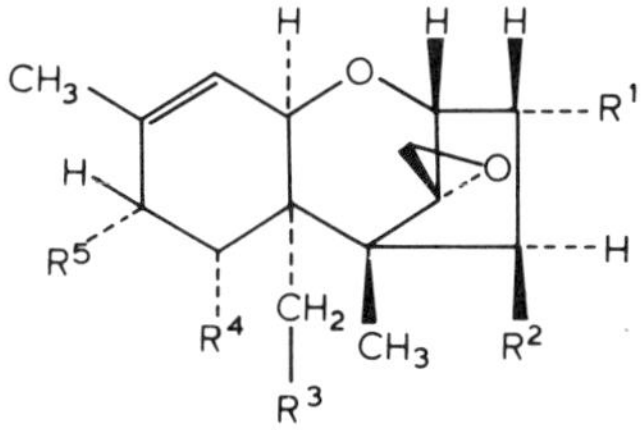

FIG. 20. Chemical structure of trichothecenes.

detecting trichothecenes at low levels would be a welcome addition to trichothecene methodology.

High-performance liquid chromatography has partly superseded thin-layer chromatography in the analysis of food for mycotoxins. The reasons for this development are obvious. Separations can be accomplished in a matter of minutes, HPLC methods generally provide good quantitative information and the equipment employed in HPLC systems can be automated rather easily. Finally, on-line coupling with a mass spectrometer has become possible, although it is too early to estimate the value of the latter technique for the determination and confirmation of mycotoxins.

HPLC has limitations as well. Although resolutions are much better than those obtained using one-dimensional TLC, the use of two-dimensional chromatography in HPLC is hardly possible. It is just the latter technique that has proved to be such a powerful separation tool when applied to thin-layer chromatography, especially when low limits of detection are required for 'dirty' sample extracts. The cost of equipment for thin-layer chromatography (except densitometers) is relatively cheap compared with the expensive instrumentation for HPLC. The extensive experience required to obtain the maximum benefit from an HPLC system constitutes another limitation, whereas TLC can be learned relatively easily. The few published studies in which HPLC has been compared with TLC for the determination of aflatoxins in peanuts,[93] and in corn and peanuts,[94] have indicated that both techniques provide results that agree rather well. In conclusion, HPLC can be an attractive alternative to TLC. However, each toxin in combination with each type of foodstuff poses a unique problem. It is the analyst's duty to find out which technique provides the best results in view of his objectives, and not to choose haphazardly a sophisticated technique for sophistication's sake.

(iv) *Gas–liquid chromatography*. The use of gas–liquid chromatography (GLC) in mycotoxin analysis has been limited, as most of the mycotoxins are not volatile and must therefore be derivatised before they can be gas chromatographed. In addition, the fact that many of the mycotoxins are readily detected and determined at low levels of concentration using TLC and HPLC techniques, as discussed in the foregoing paragraphs, has not stimulated the development of gas chromatographic assays. However, a few exceptions and some new developments in column technology justify some attention in this respect.

Although some gas chromatographic methods have become available in the 1970s for the detection and determination of patulin (see Fig. 1) in apple juice,[95] penicillic acid in maize and in rice[96] and zearalenone in maize,[97] the only significant advantage over TLC and HPLC techniques is the potential use of mass spectrometers as highly selective and sensitive detectors. However, this situation is quite different for one important group of the mycotoxins, the trichothecenes (see Fig. 20). Chemical determination of the trichothecenes by TLC and HPLC is difficult owing to the fact that these compounds have no fluorescent properties, nor do trichothecenes absorb appreciably in the UV range. Although TLC methods have been developed using visualisation reagents, the detection limits obtained are relatively high compared to GLC.

GLC permits the detection and quantitation of most of the more common trichothecenes. Bamberg[98] was the first to describe gas chromatography of T-2 toxin. Trichothecenes can be gas chromatographed as their trimethylsilyl (TMS)[99] or heptafluorobutyryl (HFB)[100] derivatives, using non-polar stationary phases such as methylsilicones (SE-30, OV-1), methyltrifluoropropylsilicone (QF-1) or methylphenylsilicone (OV-17), whereas detection occurs with flame ionisation detectors (FID) and electron capture detectors (ECD) respectively. The latter detection system is more sensitive and selective and is to be preferred. The claimed limits of detection achieved when analysing extracts of agricultural commodities with packed columns using FID vary from 10 to 1000 μg/kg, whereas those obtained with ECD vary from 0·3 to 100 μg/kg, depending on the extraction and clean-up, the type of trichothecenes and the commodity analysed.[101] These figures make it clear that any laboratory equipped with GLC instrumentation should be using ECD rather than FID for determining trichothecenes in grains.

In addition to the above-mentioned detection systems, it is also possible to couple the gas chromatograph directly to a mass spectrometer (MS). When operating in the selected ion monitoring mode, the mass spectrometer is a very selective and sensitive detector which can be used for quantitative assays of commodities for mycotoxins. GLC–MS employing selected ion monitoring has been reported not only for the determination of trichothecenes, but incidentally for a few other mycotoxins as well. Rosen and Pareles[102] and Coxon and Price[103] developed GLC–MS methods for the determination of patulin in apple juice suitable for detecting patulin at levels of contamination of *ca.* 10 μg/kg and 1 μg/kg respectively. Salhab *et al.*[104] described a GLC–MS method for quantitative determination of sterigmatocystin in grain and achieved a limit of detection of 5 μg/kg. There are few GLC–MS methods for most mycotoxins with the exception of the trichothecenes, for which such techniques are widely used. A method was described in 1976[105] using computer-controlled GLC–MS multiple ion detection to determine T-2 toxin, diacetoxyscirpenol and deoxynivalenol in feeds. For each trichothecene trimethylsilyl derivative, nine characteristic ions were monitored as peaks eluted from the GLC column. Since then a number of publications have appeared on quantitative methods for determination of trichothecenes in agricultural commodities using GLC–MS for detection or confirmation. A review of these methods with their respective limits of detection (varying from 3 to 500 μg/kg) has been given by Scott.[101] Good agreement between GLC–ECD and GLC–MS has been obtained for determinations of deoxynivalenol in naturally contaminated wheat, maize and barley.[106]

A recent development in GLC is the use of capillary columns in food analysis.[107] The advantage of the use of capillary columns over packed columns, viz. higher resolutions, has prompted the development of capillary GLC methods for trichothecenes and probably the first publication in this respect appeared in 1980.[108] Since then, several publications dealing with this subject have been published.[109–111] Bijl *et al.*[110] compared some possibilities of capillary columns coated with Ov-1701, a medium polar liquid phase, in combination with FID, ECD or MS for the detection and determination of trichothecenes. Underivatised trichothecenes (trichothecin, diacetoxyscirpenol and T-2 toxin) could be detected and quantified in extracts of maize at levels down to 50 μg/kg maize, using the combination cold on-column injection/FID detection, whereas a recovery of >80 % was obtained. Using the combination moving needle injection/ECD detection of HFB esters of the trichothecenes mentioned did not enhance selectivity and sensitivity. Capillary GLC with selected ion monitoring MS of TMS derivatives further decreased the limit of detection to *ca.* 10 μg/kg. An example of a capillary GLC separation using MS detection is given in Fig. 21.[112] A crude extract of maize has been analysed for deoxynivalenol. Deoxynivalenol was resolved as its trimethylsilylether derivative on a 35 m DB5 capillary column (0·2 mm). All components are normalised to the concentration of the most abundant species. The equivalent of 5 mg of sample (contaminated at a level of 72 mg/kg) was actually injected on the column, whose temperature was programmed from 80 to 300 °C at 15 °/min. It may be noted from Fig. 21 that the deoxynivalenol component is small relative to the extract components. However, the mass spectrometer can be focused on the characteristic ions of deoxynivalenol at the specific retention time, so that a sensitivity can be obtained which is a thousand-fold more than is actually necessary for this sample.

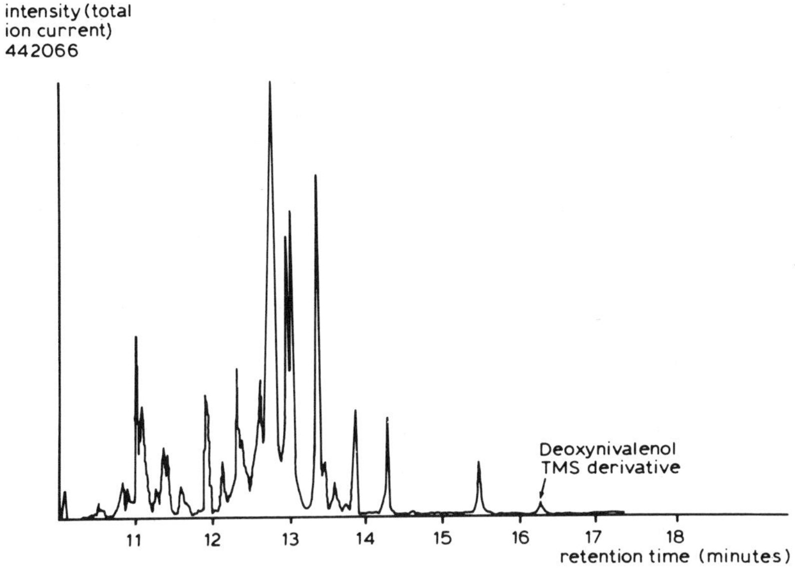

FIG. 21. Capillary GLC separation of an extract of maize, analysed for deoxynivalenol using MS detection (sample contaminated at a level of 72 mg/kg) (after Mirocha[112]).

Thanks to developments in column technology, injection systems, detection systems and advanced data retrieval systems, GLC has become a powerful technique for detection and determination of the trichothecenes. GLC–MS is a unique tool enabling precise identification. However, the practical use of this sophisticated system is limited to those laboratories who can afford to purchase these expensive computer-controlled systems.

(b) Immunochemical Procedures

Immunochemical procedures are based upon quite different principles than chromatographic procedures. An immunoassay is an assay in which the molecular recognition properties of antibodies, bio-macromolecules, are used, rather as a lock responds to a key. The key to be measured is the antigen (= the analyte). Classical immunoanalytical techniques are based on the interaction (in solution) between native antigens and specific antibodies, leading to precipitation of the antigen–antibody complex. This precipitation is a measure for the antigen or antibody concentration. These precipitation procedures are suitable for measuring antigen concentrations (in solution) in the order of μg–mg/ml. However, when the analyte is present at lower concentrations, which is often the case when mycotoxins are to be determined in foods and feeds, other more sensitive methods are needed to measure the complex formation. For these situations the antigen is labelled for the purpose of counting. The label can be a radioisotope, an enzyme, a fluorescent group or some other marker which can easily be quantified. Immunoassays are widely used in clinical analysis, where radioimmunoassay and enzyme immunoassay are the most frequently used in routine diagnostics. The number of immunoassays used in food analysis is still limited, although these techniques will become of growing importance, thanks to the rapid advances being made. The use of various types of immunochemical methods in food analysis and the background of the techniques are comprehensively discussed by Dr Daussant in Chapter 5 of this volume. For the determination of mycotoxins, the use of immunoassays has been limited to date to radioimmunoassay (RIA) and enzyme-linked immunosorbent assay (ELISA).

It is beyond the scope of this chapter to discuss the production of antibodies against mycotoxins, which generally have molecular weights too low to evoke antibodies directly when administered to animals. These so-called haptens have to be conjugated with proteins before immunisation can occur. The antibodies (antiserum) are a group of serum proteins also referred to as immunoglobulins. Most of these immunoglobulins belong to the IgG class. Because these immunoglobulins possess not only antibody reaction sites but also antigenic determinant sites, the immunoglobulins themselves can serve as antigens when injected in a foreign animal. Up to now it has been possible to evoke antibodies against aflatoxin B_1,[113] ochratoxin A,[114] T-2 toxin[115] and aflatoxin M_1.[116] Many mycotoxins have closely related chemical structures and are accordingly grouped together, e.g. the aflatoxins, ochratoxins and trichothecenes. Because of this there is, in principle, a possibility that cross-reactions could occur between antibodies evoked against a certain mycotoxin and other co-occurring toxins within the same group. In addition, cross-reactions are possible with extract components. In practice, several of the antibodies produced against mycotoxins have been shown to be rather specific; nevertheless the possibility of cross-reactivity deserves continuous attention.

For the performance of immunoassays it is further necessary to have available radiolabelled mycotoxins (for which generally ^3H is used) in the case of RIA, and enzyme-labelled mycotoxins for ELISA (for which often horseradish peroxidase is used). The latter is not a prerequisite if indirect techniques are applied, which means that use is made of enzyme-labelled anti-immunoglobulins which react with the anti-mycotoxin antibodies which, in turn, bind to the antigen (=the mycotoxin). With this technique a kind of cascade is obtained. The several variants of the RIA and ELISA systems will not be fully discussed here; only a few examples of systems used in mycotoxin research are given. For more details the reader is referred to Chapter 5 of this volume. An overview of the current state of the art of the use of immunochemical methods for the determination of mycotoxins has been published by Chu.[117]

The extraction and clean-up procedures applied in immunoassays developed for mycotoxins are generally the same as those used for chromatographic procedures, though because of the high specificity of immunoassays crude extracts are sometimes suitable. Unlike many of the extracts prepared for chromatographic techniques, the final extracts (test portions) used in immunoassays are (buffered) aqueous solutions.

In Fig. 22 the mechanism of a radioimmunoassay is outlined. The test portion, containing a known amount of labelled antigen (marked as active) and an unknown amount of unlabelled antigen (the mycotoxin looked for), is brought into contact with a fixed amount of antibody. Competition takes

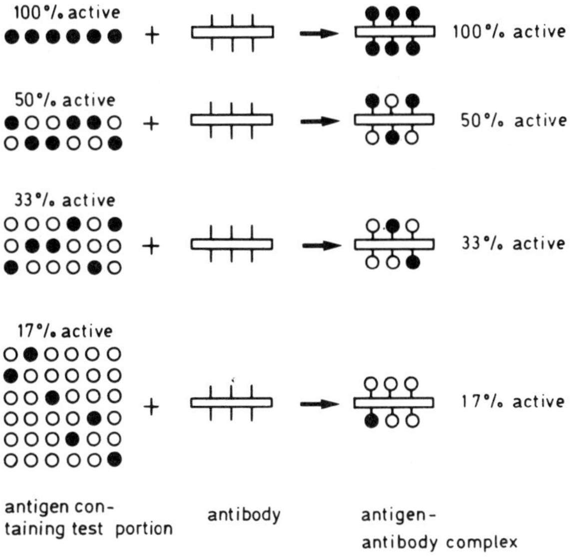

FIG. 22. Mechanism of radioimmunoassay. ●, labelled antigen; ○, unlabelled antigen.

place between labelled and unlabelled antigen for the active sites of the antibody. After a certain time equilibrium is reached, and there will remain some free antigen, the rest being bound to the antibody. The relative binding ratio of the labelled and unlabelled antigen to the antibody depends on the relative concentration ratio of the labelled and unlabelled antigen in the test portion. The lower the ratio labelled antigen/unlabelled antigen, the lower the radioactivity of the antigen–antibody complex. After separation of the antigen–antibody complex and the free fraction, the radioactivity of the complex is measured in a liquid scintillation counter. This radioactivity is a measure of the amount of unlabelled antigen (the mycotoxin looked for) in the test portion. Normally the evaluation of the amount of mycotoxin in an unknown sample is made by using a standard curve.

The first published RIA for the determination of aflatoxin B_1 has been described by Langone and Van Vunakis.[113] This method makes use of a double antibody precipitation, a technique in which a second antibody, anti-IgG, reacts specifically with the first antibody complex. In this way the difference in weight and size between the complex and the free fraction is increased, making separation easier. The procedure of Langone and Van Vunakis,[113] when applied to the analysis of extracts made from maize and peanut butter at the 1–50 μg B_1/kg level, yielded recoveries ranging from 40 to 99% for maize and 34 to 54% for peanut butter, respectively. Owing to non-specific interference, accurate results could not be obtained on samples of peanut butter that contained 1 μg aflatoxin B_1/kg. Since 1976 several RIA procedures have been developed. Sun and Chu[18] described a simple solid-phase RIA for aflatoxin B_1 which is less time-consuming than the method of Langone and Van Vunakis.[113] In this procedure the separation between the free fraction and the complex is facilitated by covalently binding the antibody to Sepharose gel. The method of Sun and Chu,[118] which is very specific for aflatoxin B_1, would allow the determination of aflatoxin B_1 in maize at levels of 5 μg/kg without the need of extensive clean-up. It was concluded that a clean-up step is necessary should the level fall below 1 μg/kg.

RIA methods have also become available for the determination of ochratoxin A,[114] aflatoxin M_1[116] and T-2 toxin.[119] In the RIA procedure for ochratoxin A, use is made of dialysis to separate free and bound toxin, whereas ammonium sulphate precipitation of the antigen–antibody complex is applied in the procedures for the determination of aflatoxin M_1 and T-2 toxin. The method for aflatoxin M_1 has a limit of detection of 0·25 μg M_1/litre milk, a limit that cannot yet compete with those obtainable with chromatographic procedures. The RIA procedure for T-2 toxin would allow the detection of 1 and 2·5 μg/kg T-2 toxin in wheat and maize samples, respectively. The fact that no adequate methods have yet been published to quantify low levels of trichothecenes in agricultural commodities, other than those making use of advanced GLC systems, makes the RIA method of Lee and Chu[119] a very interesting development which deserves further attention.

The use of enzyme-linked immunosorbent assay (ELISA) has been introduced in mycotoxin research by Lawellin *et al.*[120] for the determination of aflatoxin B_1 in urine and blood. Of the several possibilities of ELISA, the competitive assay, the titration assay (a sequential saturation variant of the competitive assay) and the inhibition assay have been applied to date for the determination of mycotoxins in foodstuffs. In Fig. 23 the mechanism of the competitive principle applied for the determination of aflatoxin B_1[121] is outlined. A polystyrene tube or

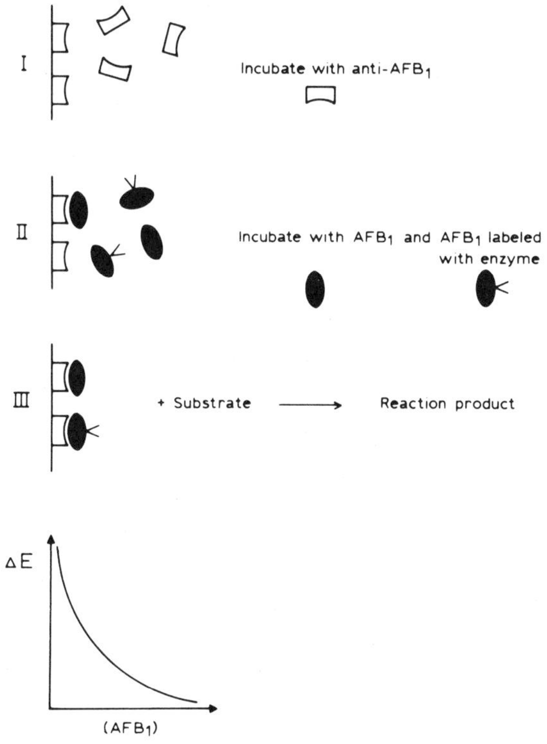

FIG. 23. Aflatoxin B_1 determination by enzyme-linked immunosorbent assay.

microtitre plate is coated with a known amount of antibody against aflatoxin B_1. After being washed, the test solution containing an unknown quantity of aflatoxin B_1 is added, followed directly by the addition of a known quantity of aflatoxin B_1 labelled with enzyme. Labelled and non-labelled aflatoxin B_1 compete for the active sites of the bound antibody. After incubation the tube or plate is washed again and the captured enzyme is determined by adding chromogenic substrate. The resulting colour can be measured instrumentally or visually. The lower the concentration of reaction product, the lower the amount of bound enzyme and the higher the aflatoxin B_1 concentration in the test portion. As with RIA, a standard curve is established from which the toxin concentration in the test portion is then determined.

The competitive assay of Pestka *et al.*,[121] in which microtitre plates are used, has been applied to the analysis of maize, wheat and peanut butter by El-Nakib *et al.*[122] It was concluded that samples containing aflatoxin B_1 at levels above 5 μg/kg can be analysed by ELISA assays directly after a simple extraction without a column chromatographic step.

A practical application of the titration assay for aflatoxin B_1 has been developed by Biermann and Terplan[123] for the determination of aflatoxin B_1 in peanut meal. In this type of ELISA separate incubation steps for aflatoxin B_1 and the enzyme conjugate are used. In the first step aflatoxin B_1 is bound to a part of the active antibody sites. In the second step the remaining active sites are back-titrated with enzyme-labelled aflatoxin B_1. Biermann and Terplan[123] reported a limit of detection of *ca.* 0·1 μg/kg and recoveries of *ca.* 90 % when samples of peanut meal, contaminated at levels of 1–10 μg/kg, were analysed. The extraction procedure described was very simple and no clean-up procedure was used.

Other mycotoxins for which ELISA procedures have been developed are aflatoxin M_1,[124] T-2 toxin[125] and ochratoxin A.[126] The limit of detection for the competitive M_1 procedure[124] was reported to be at 0·25 μg/litre of milk, which is rather high in comparison to chromatographic procedures, for which limits of detection are achievable that are *ca.* 10-fold lower.

The procedure developed for T-2 toxin,[125] also based on the competitive principle, showed limits of detection of 1 and 5 μg/kg in wheat and maize, respectively, with recoveries ranging from 95 to 98 % when the food extracts were subjected to a Sep-pak® clean-up step. These limits of detection compare favourably to those of chromatographic procedures for trichothecenes, and the ELISA procedure is one of the simpler methods to apply.

The method for ochratoxin A[126] is the first published ELISA method for mycotoxins that makes use of the inhibition principle. In this procedure, samples or ochratoxin A standards and a fixed amount of anti-ochratoxin A antisera (in excess) are added to the ochratoxin A-coated wells of a microtitre plate. The antibody that has not reacted with ochratoxin A from the samples or standards is captured by the ochratoxin A-coated inside surface of the wells. The captured antibody is a rabbit immunoglobulin. After incubation and washing, the plate is incubated with a second anti-rabbit antibody, labelled with enzyme. The captured enzyme is determined by adding chromogenic substrate. The method has been successfully applied to barley and an absolute lower limit of detection of 10 pg ochratoxin A per well has been reported. The cross-reaction of ochratoxin B was only 0·5%.

The application of immunoassays for the determination of mycotoxins is a recent development, so that collaborative studies on this subject have not yet been conducted at the time of writing. The present lack of analytical parameters derived from collaborative studies makes it difficult to estimate the value of the newcomers compared to the conventional chromatographic procedures. Incidentally, ELISA has been applied to the analysis of samples provided in the 1980 Aflatoxin Check Sample Program, sponsored by the International Agency for Research on Cancer at Lyon. It was reported[122] that the results of ELISA applied to the analysis of three samples (peanut meal, de-oiled peanut meal and yellow corn meal, naturally contaminated with aflatoxins at estimated levels of *ca.* 210, 110 and 55 μg/kg, respectively) were in excellent agreement with the overall mean values obtained by other collaborators using different methods, which offers hopeful perspectives for the wider use of ELISA.

RIA and ELISA have been compared to each other for the determination of aflatoxin B_1 in maize, wheat and peanut butter,[122] aflatoxin M_1 in milk[124] and T-2 toxin in maize and wheat.[117] It was concluded that ELISA was the preferred method for the aflatoxins. For aflatoxin B_1 more consistent data, relatively lower standard deviations and lower coefficients of variation were obtained with ELISA than with RIA. ELISA was also the preferred method for aflatoxin M_1, mainly because of its simplicity, sensitivity and selectivity. For T-2 toxin both RIA and ELISA seemed to be adequate when maize and wheat were analysed.

A major advantage of both RIA and ELISA is the possibility of complete automation, making the techniques very valuable as rapid quantitative (screening) procedures. Taking into account the disadvantages of RIA, such as the limited shelf-life activity of the radioisotopes, problems of radioactive waste disposal or licensing requirements, it is to be expected that ELISA in particular will be of growing importance as an assay technique for mycotoxins. However, a matter of continuous vigilance is the specificity of immunoassays. Although most of the antisera produced against mycotoxins seem to be quite specific, the possibility of cross-reactions cannot be fully ruled out and it is a good laboratory practice to confirm positive findings of immunoassays by methods of analysis based on other principles.

4. CONCLUSION

The present state of methodology for the determination of mycotoxins in food may be summarised as follows:

1. Bioassays may be useful in tracing sources of known and unknown mycotoxins. However, their use in the surveillance of food and foodstuffs for mycotoxins is of minor importance.
2. Chemical assays are of major importance in the determination of mycotoxins. Most widely used are those techniques which include a chromatographic step to separate the mycotoxin of interest from matrix components.
3. Mini-column chromatographic procedures are useful as screening tests for agricultural commodities if quick decisions are needed as to whether to accept or reject a lot. They have been developed mainly for aflatoxins.
4. Thin-layer chromatography, although a veteran in mycotoxin methodology, is a reliable, feasible and relatively simple separation technique with a broad field of application. Its two-dimensional application offers especially good resolution, resulting in low limits of detection.
5. High-performance liquid chromatography can be an attractive alternative to thin-layer chromatography. This more expensive technique offers the possibility of automating the ultimate separation and quantification steps. However, if time-consuming clean-up procedures are necessary, as may be the case for 'dirty' products, this advantage is limited.
6. The use of gas–liquid chromatography is limited mainly to the analysis of commodities for trichothecenes. Thanks to technical developments, gas–liquid chromatography has become a powerful technique in determining toxins of this group. Gas–liquid

TABLE 1
COMPARISON OF SOME CHARACTERISTICS OF VARIOUS CATEGORIES OF METHODS FOR THE DETERMINATION OF MYCOTOXINS IN FOOD

Category	*Scope of application*	*Reliability*	*Limits of detection*	*Equipment cost*	*Automation possible*
Bioassay	Limited	Low	High	Low	No
Mini-column	Limited	Moderate	Moderate	Low	No
TLC	Broad	High	Low	Low	No
HPLC	Broad	High	Low	High	Partly
GLC	Limited	High	Low	High	Partly
RIA	Limited	[a]	Low	High	Yes
ELISA	Limited	[a]	Low	Moderate	Yes

[a] Unknown at present.

chromatography–mass spectrometry with selected ion monitoring is currently the peak of methodological sophistication and can be used to detect and confirm the presence of mycotoxins in the μg/kg range.

7. Immunoassays such as radioimmunoassay and enzyme-linked immunosorbent assay are promising techniques. Although still in their infancy, it is to be expected that these techniques will play an important role in mycotoxin methodology in the near future.

Some characteristics of the various categories of methods discussed in this chapter are compared to each other in Table 1. It should be realised that the classification of these categories is a rather subjective matter, influenced by personal experience, opinions and preferences and therefore debatable. The final decision as to which method to apply for a certain problem may be difficult if more than one possibility exists. In such a case, it is advisable to consult the AOAC *Official Methods of Analysis* (Chap. 26) (1980)[35] and the IARC publication *Selected Methods of Analysis*, Vol. 5 (1982).[127]

ACKNOWLEDGEMENTS

The author wishes to thank J. R. Besling (Food Inspection Service, Rotterdam, The Netherlands) and G. E. Ellen, W. E. Paulsch and P. L. Schuller (National Institute of Public Health, Bilthoven, The Netherlands) for their helpful suggestions in preparing the manuscript. A. E. Buckle

(Agricultural Development and Advisory Service, Shardlow, Derby, UK) is gratefully acknowledged for correcting the manuscript. T. B. Whitaker (North Carolina State University, Raleigh, NC, USA), L. G. M. Th. Tuinstra (State Institute for Quality Control of Agricultural Products, Wageningen, The Netherlands) and C. J. Mirocha (University of Minnesota, St Paul, MN, USA) kindly provided Figs. 2–6, 19 and 21 respectively. Finally, the author would like to express his thanks to Miss Rita van Doorn for her excellent preparation of the typescript.

REFERENCES

1. Tulasne, L. R. (1853). *Ann. Sci. Nat.*, 3rd Ser., **20**, 5.
2. Sarkisov, A. H. and Kvasnina, E. S. (1948). *C.R. Acad. Sci. USSR*, **63**, 1.
3. Mayer, C. F. (1953). *Mil. Surg.*, **113**, 173.
4. Joffe, A. Z. (1965). In *Mycotoxins in Foodstuffs*, Wogan, G. N. (Ed.), MIT Press, Cambridge, Mass., p. 77.
5. Kinosita, R. and Shikata, T. (1965). In *Mycotoxins in Foodstuffs*, Wogan, G. N. (Ed.), MIT Press, Cambridge, Mass., p. 111.
6. Forgacs, J. (1962). *Feedstuffs*, **34**, 124.
7. Stevens, A. J., Saunders, C. N., Spence, J. B. and Newnham, A. G. (1960). *Vet. Rec.*, **72**, 627.
8. Asplin, F. D. and Carnaghan, R. B. A. (1961). *Vet. Rec.*, **72**, 1215.
9. Nesbitt, B. F., O'Kelly, J., Sargeant, K. and Sheridan, A. (1962). *Nature (London)*, **195**, 1062.
10. Hartley, R. D., Nesbitt, B. F. and O'Kelly, J. (1963). *Nature (London)*, **198**, 1056.
11. Allcroft, R. and Carnaghan, R. B. A. (1963). *Vet. Rec.*, **75**, 259.
12. Anon. (1980). *Environmental Health Criteria 11: Mycotoxins*, World Health Organisation, Geneva.
13. Kiermeier, F., Orth, R., Leistner, L., Eckardt, C., Spicker, G., Reiss, J. and Frank, H. K. (1981). In *Mycotoxine in Lebensmitteln*, Reiss, J. (Ed.), Gustav Fischer Verlag, Stuttgart and New York, p. 243.
14. Nieuwenhuize, J. P. van, Herber, R. F. M., Bruin, A. de, Meyer, P. B. and Duba, W. C. (1973). *T. Soc. Geneesk.*, **51**, 754.
15. Anon. (1982). *Nature (London)*, **296**, 379.
16. Mirocha, C. J., Watson, S. and Hayes, W. (1982). *Proc. Vth Int. IUPAC Symp. on Mycotoxins and Phycotoxins*, Vienna, Sept. 1982, p. 130.
17. Schuller, P. L., Stoloff, L. and Egmond, H. P. van (1983). *Proc. Int. Symp. Mycotoxins*, Cairo, Sept. 1981, p. 111.
18. Cucullu, A. F., Lee, L. S., Mayne, R. Y. and Goldblatt, L. A. (1966). *J. Am. Oil Chem. Soc.*, **43**, 89.
19. Egmond, H. P. van, Northolt, M. D. and Paulsch, W. E. (1982). *Proc. 4th Meeting on Mycotoxins in Animal Disease*, Weybridge, UK, Apr. 1981, p. 87.
20. Whitaker, T. B. (1977). *Pure and Appl. Chem.*, **49**, 1709.

21. DICKENS, J. W. (1978). In *Gesundheitsgefährdung durch Aflatoxine*, Poiger, H. (Ed.), Arbeitstagung Zürich, März 1978, Institut für Toxikologie der ETH unter der Universität Zürich, Schwerzenbach, Switzerland, p. 258.
22. DICKENS, J. W. (1977). *J. Am. Oil Chem. Soc.*, **54**, 225A.
23. WHITAKER, T. B., DICKENS, J. W., MONROE, R. J. and WISER, E. H. (1972). *J. Am. Oil Chem. Soc.*, **49**, 590.
24. WAIBEL, J. (1977). *Dtsch. Lebensm.-Rundschau*, **73**, 353.
25. DICKENS, J. W. and SATTERWHITE (1969). *Fd Technol.*, **23**, 90.
26. FRANCIS, O. J. (1979). *J. Ass. Off. Anal. Chem.*, **62**, 1182.
27. WHITAKER, T. B., DICKENS, J. W. and MONROE, R. J. (1980). *J. Am. Oil Chem. Soc.*, **57**, 269.
28. CARNAGHAN, R. B. A., HARTLEY, R. D. and O'KELLY, J. (1963). *Nature (London)*, **200**, 1101.
29. WATSON, D. H. and LINDSAY, D. G. (1982). *J. Sci. Fd Agric.*, **33**, 59.
30. BURMEISTER, H. R. and HESSELTINE, C. W. (1966). *Appl. Microbiol.*, **14**, 403.
31. CLEMENTS, N. L. (1968). *J. Ass. Off. Anal. Chem.*, **51**, 1192.
32. BROWN, R. F., WILDMAN, J. D. and EPPLEY, R. M. (1968). *J. Ass. Off. Anal. Chem.*, **51**, 905.
33. WAART, J. DE, AKEN, F. VAN and POUW, H. (1972). *Zbl. Bakt. Hyg. I. Abt. Orig.*, **222**, 96.
34. VERRET, M. J., WINBUSH, J., REYNALDO, E. F. and SCOTT, W. F. (1973). *J. Ass. Off. Anal. Chem.*, **56**, 901.
35. ANON. (1980). In *Official Methods of Analysis of the Association of Official Analytical Chemists*, Horwitz, W. (Ed.), Washington, DC, p. 414.
36. CHUNG, C. W., TRUCKSESS, M. W., GILES, A. L. and FRIEDMAN, L. (1974). *J. Ass. Off. Anal. Chem.*, **57**, 1121.
37. UMEDA, M. (1977). *In Mycotoxins in Human and Animal Health*, RODRICKS, J. V., HESSELTINE, C. W. and MEHLMAN, M. A. (Eds), Pathotox Publishers, Inc., Park Forest South, IL, p. 713.
38. UMEDA, M. (1971). *Jpn J. Exp. Med.*, **41**, 195.
39. SCHOENTAL, R. and WHITE, A. F. (1965). *Nature (London)*, **205**, 57.
40. BURMEISTER, H. R. and HESSELTINE, C. W. (1970). *Appl. Microbiol.*, **20**, 437.
41. EPPLEY, R. M. (1966). *J. Ass. Off. Anal. Chem.*, **49**, 1218.
42. HOLADAY, C. E. (1968). *J. Am. Oil Chem. Soc.*, **45**, 680.
43. ROMER, T. R. (1975). *J. Ass. Off. Anal. Chem.*, **58**, 500.
44. ROMER, T. R. and CAMPBELL, A. D. (1976). *J. Ass. Off. Anal. Chem.*, **59**, 110.
45. HOLADAY, C. E. (1976). *J. Am. Oil Chem. Soc.*, **53**, 603.
46. HOLADAY, C. E. (1980). *J. Am. Oil Chem. Soc.*, **57**, 491A.
47. ANON. (1982). *Handbook on Rapid Detection of Mycotoxins*, OECD, Paris, p. 13.
48. KIERMEIER, F. and GROLL, D. (1970). *Z. Lebensmitt.-Untersuch.*, **142**, 120.
49. STUBBLEFIELD, R. D. (1979). *J. Ass. Off. Anal. Chem.*, **62**, 201.
50. PATTERSON, D. S. P. and ROBERTS, B. A. (1979). *J. Ass. Off. Anal. Chem.*, **62**, 1265.
51. EGMOND, H. P. VAN, PAULSCH, W. E., DEIJLL, E. and SCHULLER, P. L. (1980). *J. Ass. Off. Anal. Chem.*, **63**, 110.
52. STUBBLEFIELD, R. D. and SHOTWELL, O. L. (1981). *J. Ass. Off. Anal. Chem.*, **64**, 964.

53. Paulsch, W. E., Egmond, H. P. van and Schuller, P. L. (1982). *Proc. Vth Int. IUPAC Symp. Mycotoxins and Phycotoxins*, Vienna, Sept. 1982, p. 40.
54. Guiochon, G., Gounord, M. F., Siouffi, A. and Zakaria, M. (1982). *J. Chromatogr.*, **250**, 1.
55. Anon. (1976). *Off. J. Eur. Comm.*, L 102.
56. Beljaars, P. R., Verhülsdonk, C. A. H., Paulsch, W. E. and Liem, D. H. (1973). *J. Ass. Off. Anal. Chem.*, **56**, 1444.
57. Jemmali, M. (1977). *Arch. Inst. Pasteur Tunis*, **3–4**, 249.
58. Scott, P. M., Walbeek, W. van, Kennedy, B. and Anyeti, D. (1972). *J. Agric. Fd Chem.*, **20**, 1103.
59. Northolt, M. D., Egmond, H. P. van, Soentoro, P. S. S. and Deijll, W. E. (1980). *J. Ass. Off. Anal. Chem.*, **63**, 115.
60. Stack, M. and Rodricks, J. V. (1971). *J. Ass. Off. Anal. Chem.*, **54**, 86.
61. Smith, R. H. and McKerman, W. (1962). *Nature* (*London*), **195**, 1301.
62. Andrellos, P. J. and Reid, G. R. (1964). *J. Ass. Off. Anal. Chem.*, **47**, 801.
63. Pohland, A. E., Cushmac, M. E. and Andrellos, P. J. (1968), *J. Ass. Off. Anal. Chem.*, **51**, 907.
64. Pohland, A. E., Yin, L. and Dantzman, J. G. (1970). *J. Ass. Off. Anal. Chem.*, **53**, 101.
65. Przybylski, W. (1975). *J. Ass. Off. Anal. Chem.*, **58**, 163.
66. Verhülsdonk, C. A. H., Schuller, P. L. and Paulsch, W. E. (1977). *Zeszyty Problemowe Postępów Nauk Rolniczych*, **189**, 277.
67. Trucksess, M. (1976). *J. Ass. Off. Anal. Chem.*, **59**, 722.
68. Egmond, H. P. van, Paulsch, W. E. and Schuller, P. L. (1978). *J. Ass. Off. Anal. Chem.*, **61**, 809.
69. Egmond, H. P. van and Stubblefield, R. D. (1981). *J. Ass. Off. Anal. Chem.*, **64**, 152.
70. Kleinau, G. (1981). *Die Nahrung*, **25**, K9.
71. Seiber, J. N. and Hsieh, D. P. H. (1973), *J. Ass. Off. Anal. Chem.*, **56**, 827.
72. Pons, W. A. (1976). *J. Ass. Off. Anal. Chem.*, **59**, 101.
73. Kmieciak, S. (1976). *Z. Lebensmitt. Unters.-Forsch.*, **160**, 321.
74. Takahashi, D. M. (1977). *J. Chromatogr.*, **131**, 147.
75. Haghighi, B., Thorpe, C. W., Pohland, A. E. and Barnett, R. (1981). *J. Chromatogr.*, **206**, 101.
76. Panalaks, T. and Scott, P. M. (1977). *J. Ass. Off. Anal. Chem.*, **60**, 583.
77. Zimmerli, B. (1977). *J. Chromatogr.*, **131**, 458.
78. Manabe, M., Goto, T. and Maisuura, S. (1978). *Agric. Biol. Chem.*, **42**, 2003.
79. Davis, N. D. and Diener, U. L. (1979). *J. Appl. Biochem.*, **1**, 123.
80. Davis, N. D. and Diener, U. L. (1980). *J. Ass. Off. Anal. Chem.*, **63**, 107.
81. Thorpe, C. W., Ware, G. M. and Pohland, A. E. (1982). *Proc. Vth Int. IUPAC Symp. Mycotoxins and Phycotoxins*, Vienna, Sept. 1982, p. 52.
82. Diebold, G. J. and Zare, R. N. (1977). *Science*, **196**, 1439.
83. Glancy, E. M. (1978). *Proc. 3rd Meeting on Mycotoxins in Animal Disease*, Weybridge, UK, Apr. 1978, p. 28.
84. Fremy, J. M. and Boursier, B. (1981). *J. Chromatogr.*, **219**, 156.
85. Tuinstra, L. G. M. Th. and Haasnoot, W. (1982). *Fresenius Z. Anal. Chem.*, **312**, 622.

86. WEI, R. D., CHANG, S. C. and LEE, S. S. (1980). *J. Ass. Off. Anal. Chem.*, **63**, 1269.
87. COHEN, H. and LAPOINTE, M. (1981). *J. Ass. Off. Anal. Chem.*, **64**, 1372.
88. WINTERLIN, W., HALL, G. and HSIEH, D. P. H. (1979). *Anal. Chem.*, **51**, 1873.
89. HUNT, D. C., PHILP, L. A. and CROSBY, N. T. (1979). *Analyst*, **104**, 1171.
90. HUNT, D. C., MCCONNIE, B. R. and CROSBY, N. T. (1980). *Analyst*, **105**, 89.
91. CROSBY, N. T. (1982). *Proc. 4th Meeting on Mycotoxins in Animal Disease*, Weybridge, UK, Apr. 1981, p. 70.
92. COOPER, S. J., NORTON, D. M. and CHAPMAN, W. B. (1982). *Proc. 4th Meeting on Mycotoxins in Animal Disease*, Weybridge, UK, Apr. 1981, p. 63.
93. CROSBY, N. T. and HUNT, D. C. (1978). *Proc. 3rd Meeting on Mycotoxins in Animal Disease*, Weybridge, UK, Apr. 1978, p. 34.
94. VRIES, J. W. DE and CHANG, H. L. (1982). *J. Ass. Off. Anal. Chem.*, **65**, 206.
95. POHLAND, A. E., SANDERS, K. and THORPE, C. W. (1970). *J. Ass. Off. Anal. Chem.*, **53**, 692.
96. PERO, R. W., HARVAN, D., OWENS, R. G. and SNOW, J. P. (1972). *J. Chromatogr.*, **65**, 501.
97. MIROCHA, C. J., SCHAUERHAMER, B. and PATHRE, S. V. (1974). *J. Ass. Off. Anal. Chem.*, **57**, 1104.
98. BAMBERG, J. R. (1969). Ph.D. Thesis, University of Wisconsin, Madison.
99. PATHRE, S. V. and MIROCHA, C. J. (1977). In *Mycotoxins in Human and Animal Health*, Rodricks, J. V., Hesseltine, C. W. and Mehlman, M. A. (Eds), Pathotox Publishers, Inc., Illinois, p. 713.
100. ROMER, T. R., BOLING, T. M. and MCDONALD, J. L. (1978). *J. Ass. Off. Anal. Chem.*, **61**, 801.
101. SCOTT, P. M. (1982). *J. Ass. Off. Anal. Chem.*, **65**, 876.
102. ROSEN, J. D. and PARELES, S. R. (1974). *J. Agric. Fd Chem.*, **22**, 1024.
103. COXON, D. T. and PRICE, K. R. (1978). *Proc. 3rd Meeting on Mycotoxins in Animal Disease*, Weybridge, UK, Apr. 1981, p. 23.
104. SALHAB, A. S., RUSSEL, G. F., COUGHLIN, J. R. and HSIEH, D. P. H. (1976). *J. Ass. Off. Anal. Chem.*, **59**, 1037.
105. MIROCHA, C. J., PATHRE, S. V., SCHAUERHAMER, B. and CHRISTENSEN, C. M. (1976). *Appl. Environ. Microbiol.*, **32**, 553.
106. SCOTT, P. M., LAU, P. Y. and KANHERE, S. R. (1981). *J. Ass. Off. Anal. Chem.*, **64**, 1364.
107. MANNING, D. J. (1978). In *Developments in Food Analysis Techniques—1*, King, R. D. (Ed.), Applied Science Publishers, London, p. 155.
108. SZATHMÁRY, CS., GALÁCZ, J., VIDA, L. and ALEXANDER, G. (1980). *J. Chromatogr.*, **191**, 327.
109. COHEN, H. and LAPOINTE, M. (1982). *J. Ass. Off. Anal. Chem.*, **65**, 1429.
110. BIJL, J., ROELENBOSCH, M. VAN, SANDRA, P., SCHELFAUT, M. and PETEGHEM, C. VAN (1982). *Proc. Vth Int. IUPAC Symp. Mycotoxins and Phycotoxins*, Vienna, Sept. 1982, p. 12.
111. LÁSZTITY, R., VÁNYI, A. and BATA, A. (1982). *Proc. Vth Int. IUPAC Symp. Mycotoxins and Phycotoxins*, Vienna, Sept. 1982, p. 28.
112. MIROCHA, C. J. (1983). Personal communication.
113. LANGONE, J. J. and VUNAKIS, H. VAN (1976). *J. Natl Cancer Inst.*, **56**, 591.

114. Chu, F. S., Chang, F. C. C. and Hinsdill, R. D. (1976). *Appl. Environ. Microbiol.*, **31**, 831.
115. Chu, F. S., Grossman, S., Wei, R. D. and Mirocha, C. J. (1979). *Appl. Environ. Microbiol.*, **37**, 104.
116. Harder, W. O. and Chu, F. S. (1979). *Experientia*, **35**, 1104.
117. Chu, F. S. (1983). *Proc. Int. Symp. Mycotoxins*, Cairo, Sept. 1981, p. 177.
118. Sun, P. S. and Chu, F. S. (1977). *J. Fd Safety*, **1**, 67.
119. Lee, S. and Chu, F. S. (1981). *J. Ass. Off. Anal. Chem.*, **64**, 156.
120. Lawellin, D. W., Grant, D. W. and Joyce, B. K. (1977). *Appl. Environ. Microbiol.*, **34**, 94.
121. Pestka, J. J., Gaur, P. K. and Chu, F. S. (1980). *Appl. Environ. Microbiol.*, **40**, 1027.
122. El-Nakib, O., Pestka, J. J. and Chu, F. S. (1981). *J. Ass. Off. Anal. Chem.*, **64**, 1077.
123. Biermann, A. and Terplan, G. (1981). *Arch. Lebensmitt. Hyg.*, **33**, 17.
124. Pestka, J. J., Li, Y., Harder, W. O. and Chu, F. S. (1981). *J. Ass. Off. Anal. Chem.*, **64**, 294.
125. Pestka, J. J., Lee, S. S., Lau, H. P. and Chu, F. S. (1981). *J. Am. Oil Chem. Soc.*, **58**, 940A.
126. Morgan, M. R. A., Matthew, J. A., McNerney, R. and Chan, H. W. S. (1982). *Proc. Vth Int. IUPAC Symp. Mycotoxins and Phycotoxins*, Vienna, Sept. 1982, p. 32.
127. Egan, H. (1982). *Environmental Carcinogens: Selected Methods of Analysis*, Vol. 5: *Some Mycotoxins*, IARC Scientific publ. No. 44, Lyon.

Chapter 4

FOOD AND ITS PESTICIDES

R. W. YOUNG
Virginia Polytechnic Institute and State University, Blacksburg, USA

1. INTRODUCTION

Our main objective is the analysis of pesticide groups (fungicides, herbicides, insecticides and miticides). These pesticides are chemically grouped as chlorinated hydrocarbons, organic phosphates, carbamates and chlorophenoxy compounds, which are often found in foods. We have indicated analytical methods, based on existing modified multi-residues,[1,2] which will detect numerous pesticides simultaneously, rather than the detection of only a single residue, thus saving time as well as money. Only in special cases are the multi-residue procedures not suitable, requiring the use of specific residue procedures for the analysis of a single pesticide. The majority of pesticides used are chlorinated hydrocarbons, organic phosphates and carbamates which are included here with modifications for carbamates in one multi-residue procedure, while the chlorophenoxy herbicides make up a second multi-residue grouping.

Although our primary interest is in the area of pesticides, two major non-pesticide industrial organic compounds, polychlorinated biphenyls and phthalates, are sometimes found in food products as contaminants along with pesticide residues, making it essential that a means for identification be established, as they are often mistaken for pesticides. The polychlorinated biphenyls, better known as PCBs or Aroclors 1248, 1254, 1260, etc., are named depending upon the number of chlorine atoms present. The PCBs give numerous peaks upon separation on gas chromatography (GC), and the separated constituents often have the same retention time as various pesticides. They may enter food products by contamination from plastics, cardboard, paints, other manufactured products, or from electrical systems. Phthalates are used as plasticisers in plastics and are released into water, soil and air as the plastics break down. Plastics may also contaminate food when packaging on food products breaks down due to overheating.

Water supplies may also be contaminated by pesticides, so the analysis of water has also to be considered. Pesticides in water can also be transferred to food by the use of contaminated water during either food production or processing.

When working with foods, we are not only looking at various pesticides requiring different extraction conditions, clean-up and detection procedures, but we are also sampling a variety of food types that require different sample preparation and analysis, including fatty (fat), oily (peanuts), waxy (apples), non-fatty (lettuce), soft (ripe tomatoes) or hard and dry (popcorn).

Included in a successful analytical procedure for a quantitatively accurate and precise analysis of pesticides is the influence of the analyst. It is the analyst who holds the key to high-quality data, including an understanding of equipment performance, recognising poor performance and reacting positively to keep the operation performing reliably. Only by the willingness of the analytical chemist to devote painstaking care to the operation can consistently reliable results be obtained.[3]

2. GENERAL DESCRIPTION OF PESTICIDE RESIDUE ANALYTICAL PROCEDURES

The analysis of pesticide residues can be outlined as follows:

1. Sampling and preparation.
2. Extraction from a sample matrix.
3. Clean-up: removal of interfering co-extractives.
4. Identification and estimation of quantity.
5. Confirmation of pesticides.

The analytical methods are important to ensure that the analyst has been provided with a reasonable set of data. However, a number of other factors must be considered to ensure that the data have been properly interpreted to solve the question under consideration.

If a representative sample has not been obtained, the data are meaningless in respect to the problem. To compare data, either between samples from the same area or from areas with environmental differences, variations such as soil type, temperature and light intensity must be considered. In the case of plant and animal tissues, the selection of tissue type should be carefully undertaken and should not vary between samples of the same type. Many examples are given in the contents of the methods. If it is found that a large number or large concentration of interfering compounds is present, the sample size should be kept as small as possible.

In obtaining the samples, extreme care must be used in procuring chemically clean collection containers. These containers should be glass or plastic. As soon as possible after collection, the extraction of the residues from the sample should begin. If the analyst is unable to extract residues within a reasonable length of time, the sample should either be refrigerated or frozen. In some cases it may be better to finely chop or grind the sample before freezing.

Extraction of residues may be performed using a variety of methods, depending upon the type of sample matrix. Examples of various sample matrices that are encountered may be dealt with as follows. Water may be extracted by the addition of an organic solvent accompanied by stirring with a magnetic stirrer, followed by separation of the solvent from the water, and subsequent clean-up, if required, before detection by gas chromatography. Food items may be extracted as given in Section 5.1, using blender or Polytron techniques.

Clean-up procedures are determined by the types of contamination present such as fat, wax, oil or industrial pollutants. The clean-up may be performed using a column (Florisil, silicic acid, silica gel, carbon or other suitable materials), or by partitioning. These column clean-up procedures not only separate the residues from interfering compounds, but may be used to separate residues from one another by using different solvents or solvent concentrations, hence eluting them in different fractions. After eluting, the pesticide residues in the different fractions may then be identified using GC. This could not otherwise be accomplished, as many of the residues have the same retention time on some GC columns, thus escaping detection.

Identification and estimation of the residues found in food are accomplished with GC, by use of different column packings and detectors. The ^{63}Ni electron capture detector has the highest degree of sensitivity, and is capable of detecting most organic residues, but cannot distinguish between residues containing nitrogen, chlorine, sulphur and/or phosphorus. To aid in the detection as well as confirmation of a certain residue, other detectors are used. Although the sensitivity of these detectors is lower than the electron capture detector, and the sample must be concentrated or a larger sample used, these detectors are specific to certain types of pesticides. These detectors include the hydrogen flame ionisation detector (which will detect any volatile organic compound), the flame photometric detector (which is specific for organosulphur or organophosphorus compounds), the Hall detector (for chlorine and nitrogen) and the N–P detector (for nitrogen or phosphorus). Mass spectrophotometry equipment may be used to confirm the identity of residues for which the positive identification is in doubt. However, few laboratories are currently equipped with this instrument.

The majority of the procedures used within the text have been modified from methods provided by the US Environmental Protection Agency[4] and by private correspondence. Only by years of laboratory experience may one obtain a working knowledge of pesticide analysis. Only by the close cooperation between residue groups to standardise procedures, as to sampling, extraction, and clean-up techniques and identification of residues, can the data become usable in solving residue problems in foods.

3. LABORATORY REQUIREMENTS FOR PESTICIDE ANALYSIS

3.1. Reagents

The sensitivity required for the analysis of pesticides ranges from micrograms for colorimetric analysis, to parts per billion (ppb) for gas chromatography, requiring the use of very high purity solvents. Solvents and other reagents used in pesticide analysis may be purchased, if you do not intend to redistill them in glass, bearing the manufacturer's designation of 'pesticide quality', 'pesticide organic grade', 'Resi-analysed' or 'nanograde, distilled in glass'. Upon purchasing, each lot must be checked for assurance of freedom from any impurity that may have escaped the manufacturer's quality control. The impurity may cause degradation of the pesticide or may interfere in the determinative step. Alternatively, solvents may be purchased as technical or commercial grade and distilled in an all-glass still in the laboratory.

Solvents used in pesticide residue analysis include acetone, acetonitrile, 95% ethanol, benzene, chloroform, ethyl acetate, ethyl ether (peroxide-free), hexane, isopropanol, methanol, methylene chloride, petroleum ether and toluene.

3.2. Other Chemicals, Special Reagents and Column Packings for Clean-up

All materials used such as sodium sulphate, sodium chloride and glass wool must be chemically cleaned by washing with nanograde solvents. Included in the following list are column packings used for cleaning and separating pesticides from a food sample: Celite 545 (Johns-Mansville, Denver, CO); charcoal, Norit-SG-Extra (American Norit Co., Inc., Jacksonville, FL); charcoal, Nuchar. S-N, silicic acid, 325 mesh, cotton, absorbent (Fisher Scientific Co., Inc., Raleigh, NC); dimethylchlorosilane (Pierce Chemicals Co., Rockford, IL); Florisil, PR Grade, 60–80 mesh, heated at 1250 °F (677 °C) for 3 h to activate (Floridan Co., Berkeley Springs, WV); and the Sep-pak Silica cartridge, normal phase (Waters Associates, Milford, MA).

Florisil used for cleaning up pesticide samples requires special handling. Upon receipt, Florisil must be stored in brown glass bottles with aluminium foil-lined screw caps. To ensure that a supply of dry Florisil is available for immediate use in sample clean-up, or ready for making columns requiring a certain percentage of moisture for separation of certain pesticides from one another, a bottle of Florisil should be kept in a forced draught oven at 130 °C. Each lot of Florisil purchased must be tested to determine the amount of column packing required. Adsorption capacity is based on the adsorption of lauric acid on Florisil. An excess of lauric acid[5] is added to 2 g of Florisil, which is heated overnight at 130 °C in a 25 ml Erlenmeyer flask. Remove the Florisil from the oven, stopper, cool to room temperature, add 20 ml of lauric acid solution (concentration 400 mg, prepared by adding 5 g lauric acid to 250 ml volume in hexane, giving 20 mg/ml), stopper, shake for 15 min, allow to settle, and pipette 10 ml of the supernatant into a 125 ml Erlenmeyer flask. Add 50 ml neutral absolute alcohol (adjusted with 0·05 N NaOH to pH 7·0), plus 3 drops of phenolphthalein indicator, and titrate with 0·05 N NaOH to the end point (pink). The weight adsorbed is used to calculate the proportion equivalent quantities of Florisil:

$$\text{Equivalent quantity of Florisil batch required for a column} = \frac{110 \times 20\,\text{g}}{\text{Lauric acid value of batch}}$$

$$\begin{aligned}\text{Lauric acid value} &= \text{mg lauric acid/g Florisil} \\ &= 200 - (\text{ml required for titration} \\ &\qquad \times \text{mg lauric acid/ml } 0{\cdot}05\,\text{N NaOH})\end{aligned}$$

Example:

$$\text{Lauric acid value} = 200 - \left(\frac{3\,\text{ml} \times 200\,\text{mg}}{20\,\text{ml}}\right)$$
$$= 200 - \left(\frac{600}{20}\right)$$
$$= 200 - 30$$
$$= 170$$
$$\text{Equivalent quantity of Florisil to use} = \frac{110 \times 20\,\text{g}}{170}$$
$$= 0{\cdot}6471 \times 20$$
$$= 12{\cdot}94\,\text{g of Florisil required per column}$$

To standardise Florisil, a mixture of dieldrin, malathion and azinphosmethyl is added to a 2·5 cm i.d. column with 13 g of deactivated Florisil. The Florisil column should be made up just prior to use. The column is eluted successively with 300 ml portions of 30 % methylene chloride in hexane, 10 % ethyl acetate in hexane and 30 % ethyl acetate in hexane. Dieldrin should be eluted in the first fraction, malathion in the second and azinphosmethyl in the third fraction, with all recoveries being greater than 90 %. Late elution of malathion indicates insufficient deactivation, and the need for more polar solvents (increase in percentage of methylene chloride or ethyl acetate). Early elution indicates over-deactivation, requiring less polar solvents for chromatography (lower percentage of methylene chloride or ethyl acetate).

If pesticides suspected to be present cannot be separated on Florisil, they can often be separated on a silicic acid column. For example, a silicic acid column chromatographic procedure can be used to separate DDT and its analogues from PCBs.[6,7]

4. PREPARATION OF STANDARD SOLUTIONS FOR PESTICIDES

Obtaining quantitative determinations of pesticides for gas chromatography requires that standards be accurate at suitable concentrations required for the pesticide or pesticide groups under investigation. Standards should be purchased from a reliable chemical supply house in quantities of 200 mg and stored under refrigeration in the dark. The concentrated standards of chlorinated compounds and triazines should maintain uniform strength for a 12-month period at -10 to $-15\,°C$, while the organophosphate standards are less stable, and should be held no longer than 6 months at $-15\,°C$.

Solutions used as standards are generally made up as 'stock standard solutions', which in turn are diluted and used as 'working standard solutions'. Nanograde solvents should be used in preparing 'stock or working standard solutions'. Hexane, benzene, 1 % methanol in benzene, isooctane or other suitable solvents are normally used. Although isooctane is favoured as the solvent for the working standard, since the many repeated bottle openings greatly increase the evaporation, it subsequently maintains the concentration of standards more constant than if a lower boiling point solvent is used. However, it is really not used to any great extent. In some instances the sensitivity of the pesticide 'Working Stock Standard Solution C' is increased, if made up in 1 % methanol in benzene.

4.1. Stock Standard Solutions

Generally the concentration of pesticide standards is prepared to meet the response of the pesticide and the detector used for the detection on the GC unit. A less concentrated standard is required for the electron capture than for the hydrogen flame or the flame photometric detector.

The final data depend not only upon the analytical procedure, but also on the careful weighing and preparation of the standard. It is necessary that extreme care be used in the preparation of the standard. If an error is made here, all subsequent dilutions for the life of the standard will be inaccurate. Obviously, all quantitation of samples will be similarly incorrect. The standard levels given below are the best dilutions for use on the electron capture detector.

'Stock Standard Solution A' is the first standard made in the series which requires weighing the pesticide chemical to an accuracy of 0·1 mg. Weigh 100 mg of pesticide standard, transfer with a suitable solvent to a 100 ml volumetric flask (1 mg/ml) or to a size of flask meeting your requirements, make the standard to volume, stopper and shake well. The larger the amount of standard weighed, the less the error.

'Stock Standard Solution B' is prepared by pipetting 1 ml of 'Stock Standard Solution A' to a 100 ml volumetric flask (10 μg/ml), or an amount suitable for your requirements, add the proper solvent, stopper and shake.

'Working Standard Solution C' is prepared by diluting 1 ml of 'Stock Standard Solution B' to 10 ml in a volumetric flask (1 μg/ml or 1 ng/μl), using 1 % methanol in benzene or a suitable solvent. Another concentration often used as a 'Working Standard Solution C' is 0·1 ml of 'Stock Standard Solution B' made up to 10 ml in a volumetric flask (0·01 μg/ml or 0·01 ng/μl).

The dilutions given are the standard levels found most suitable for the working requirements for the ^{63}Ni electron capture detector. Of course, the pesticide being analysed and the column packing used in the GC unit determine the final standard concentration selected by the analyst.

Standards, as well as standard solutions, should always be stored in an explosion-proof refrigerator. If evaporation of solvents occurs in a conventional refrigerator, sparks from electrical circuits could cause an explosion.

4.2. Equipment and Sample Preparation

Since the level of detection ranges down to 1 μg/litre (ppb) or less, extreme care must be taken to obtain residue-free collection containers. Scrupulously clean glassware and equipment in the laboratory as well as the use of nanograde chemicals is mandatory.

Regardless of the type of pesticide sample collected, it should be collected in glass or a plastic bag. Plastic bags must be pretested to ensure that they are residue-free. All glassware, such as sample bottles, columns, Erlenmeyer flasks, beakers, separatory funnels (only with Teflon stopcocks), cylinders, Buchner funnels, centrifuge tubes, French bottles, and equipment such as blenders, Omni-Mixer containers and Polytron probes, are cleaned as follows. Glassware and equipment should be rinsed in tap water and then isopropanol, washed in hot water with a phosphorus-free detergent, followed by 4 flowing tap water rinses, 4 flowing distilled water rinses, a nanograde acetone rinse and dried in a forced draught oven.

Care must be taken to avoid cardboard and paper containers, or paper liners in jar lids, as they may adsorb the pesticides from the sample, or may contain chemicals that may interfere with the determination of pesticides. Also, plastic containers such as Tygon or similar materials may contain plasticisers or PCBs which interfere with GC determination. As an example, Tygon tubing, used as a hose on a laboratory faucet, as it ages becomes white, hardened and precipitates the leaching of plasticisers, which may result in a number of large peaks on the GC scan. If Tygon tubing is used as a hose in this manner, it should be replaced monthly.

Upon receiving the sample (fresh or processed vegetable, fruit, meat or water) a representative portion should be taken, finely chopped, mixed, analysed immediately, or frozen (either whole or chopped) until a suitable time for analysis. Another means of preparing the sample for analysis would be the immediate extraction of the chopped sample into the appropriate solvent and storing of the sample in a dark explosion-proof refrigerator until analysis can be resumed.

5. MULTI-RESIDUE PROCEDURES FOR PESTICIDE RESIDUES IN PLANT TISSUE USED AS FOOD PRODUCTS

5.1. Chlorinated Hydrocarbons and Organic Phosphates

Plant materials being investigated for pesticide residues should be obtained in sufficient quantities to supply a representative sample for the analysis, as well as for the possible need to determine more than one group of residues, especially if the residue present is unknown, or if additional re-runs of the sample are required. If it is suspected that a number of pesticide residues may be present, more than one procedure may be required for the investigation. Also, the various types of food being analysed require different methods of laboratory preparation, which may vary from sample to sample. The composition of plant tissue ranges in texture from the succulence of leaf lettuce or ripe tomatoes to the hardness of popcorn.

The plant sample may be chopped in a Hobart food cutter (leaf lettuce), blended in a Waring Blender (tomato) or ground in a Wiley Mill (popcorn). After the sample has been ground, it is mixed, a representative sample taken and either analysed immediately or packaged in a chemically clean plastic bag and frozen until it is convenient to perform the analysis. Samples should be analysed within 2 months, otherwise breakdown of residues may have occurred.

A 25–50 g finely chopped or ground sample is weighed, transferred to a blender jar, Omni-Mixer container or a French square bottle for the Polytron, 200 ml of 65 % acetonitrile in distilled water is added and the sample is homogenised at high speed for 2 min. Filter the blended sample through a Schleicher & Schuell No. 588 fluted filter paper into a 1000 ml separatory funnel. Add 100 ml of 50 % ethyl ether in petroleum ether to the separatory funnel, stopper and shake vigorously for 1 min, venting as necessary with the stopcock to reduce pressure. Add 600 ml of distilled water and approximately 5 g of sodium chloride (NaCl), stopper, shake and allow solvent layers to separate for 10 min.

Drain the aqueous layer into a second 1000 ml separatory funnel, add 100 ml of 50 % ethyl ether in petroleum ether to the second separatory funnel, stopper and shake vigorously for 1 min, vent as necessary. Let stand for 10 min for layers to separate and discard the aqueous layer.

Combine the ether layers in the first separatory funnel, rinsing the second separatory funnel with small amounts of petroleum ether and adding to the first separatory funnel. Add 10 ml of saturated NaCl solution and 100 ml of distilled water to the ether, stopper, *shake gently* for 30 s, venting as necessary, and allow 10 min for separation of layers.

Discard aqueous layer and drain the ether layer through a 4 cm column (22 mm i.d.) of anhydrous sodium sulphate (Na_2SO_4) (to remove any traces of water) into a 400 ml beaker, placing the beaker in a warm water bath (40 °C), and evaporate on an airflow evaporator, or collect the sample in a 500 ml Erlenmeyer flask and evaporate on a flash evaporator, *just to dryness*. If the sample should remain dry for even a short period, a large percentage of the pesticides may be lost.

At this stage in the analysis it can be determined whether or not a partitioning step is necessary to clean up the sample. If the sample container has a noticeable layer of fatty or oily material, the sample must be partitioned before continuing with a column clean-up step (the partitioning procedure is outlined below).

5.1.1. Partitioning and Clean-up Procedure for Oily Plant Products (Corn, Peanuts, Peanut Butter, etc.)

If a noticeable oily layer from extracted food materials is found in the sample beaker after extraction, the oily layer is washed using 30 ml of petroleum ether into a 125 ml separatory funnel (containing 30 ml of acetonitrile saturated with petroleum ether), stoppered and shaken on a mechanical shaker for 5 min, then allow 10 min for the layers to separate.

Drain the acetonitrile layer (bottom layer) into a 1000 ml separatory funnel and save. Similarly extract the petroleum ether layer in the 125 ml separatory funnel 3 more times with 30 ml portions of acetonitrile saturated with petroleum ether, saving the acetonitrile layer each time as above. Discard the petroleum ether layer after the final shaking.

Add 100 ml petroleum ether to the combined acetonitrile extracts in the 1000 ml separatory funnel. Stopper and shake vigorously for 1 min, venting, as needed, with the stopcock to release any pressure.

Add 600 ml of distilled water and 5 g of NaCl. Stopper, shake vigorously for 1 min, and allow 10 min for separation of the solvent layers.

Drain the aqueous layer into a second 1000 ml separatory funnel. Add 100 ml of petroleum ether to the second separatory funnel containing the aqueous layer. Stopper and shake vigorously for 1 min, venting as necessary. Allow to stand and separate layers for 10 min.

Discard the aqueous layer and combine the ether layers in the first separatory funnel. Use several small rinsings to clean the second funnel with petroleum ether, adding the rinsings to the first separatory funnel. Add 10 ml of saturated NaCl solution and 100 ml of distilled water to the petroleum ether. Stopper and shake *gently for 30 s*, vent as necessary and allow 10 min for separation.

Discard the aqueous layer, and drain the ether layer into a 500 ml Erlenmeyer flask through a glass column (22 mm i.d.) with 4 cm of anhydrous Na_2SO_4, to remove any traces of water.

Evaporate the sample on a rotary flash evaporator to approximately 5–10 ml of petroleum ether and continue with the Florisil column or other column clean-up procedures.

5.1.2. COLUMN CLEAN-UP: FLORISIL

Prepare the Florisil column just prior to use. Using a 250 mm × 22 mm i.d. glass chromatographic column, with a 250 ml bulb reservoir and a Teflon stopcock (glass stopcocks cannot be used as their lubrication grease interferes with detection of residues), prepare the Florisil column as follows. Insert a glass wool plug (6 mm) in the column, followed by a 4 mm layer of anhydrous Na_2SO_4, 13 g (or appropriate amount, as determined by the lauric acid test) of activated Florisil (store Florisil in an oven at 130 °C until ready to use) and cover with a 6 mm layer of anhydrous Na_2SO_4. As the contents of the column are added, tap the sides with a wooden stick or vibrator to pack the column evenly.

Prewash the column with 50–100 ml petroleum ether, discard the prewash and place a clean collection flask under the column. At no time in the entire column run should the level of the solvent be allowed to go below the level of the Na_2SO_4 on top of the column.

Add the sample quantitatively to the column using 10–15 ml of petroleum ether to rinse the sample from the sample container on to the column, follow with 130 ml of 6 % ethyl ether (with 2 % ethanol) in petroleum ether (the 2 % ethanol in ethyl ether prevents formation of peroxides) and adjust the flow rate to 2–3 ml/min. This represents fraction I. Pesticide residues eluted in fraction I are indicated in Table 1.

Just before fraction I has drained completely into the Florisil, change collection flask and immediately add 130 ml of 15 % ethyl ether (with 2 % ethanol) in petroleum ether. This is fraction II. Maintain the same flow rate as used in fraction I. If necessary, a third fraction can be procured by using 130 ml of 50 % ethyl ether (with 2 % ethanol) in petroleum ether.

After the Florisil column has drained, evaporate fractions I, II and III *just to dryness* in an airflow or flash evaporator and immediately transfer

TABLE 1
FLORISIL ELUTION PATTERNS

Fraction I (*6% ethyl ether in petroleum ether*)	*Fraction II* (*15% ethyl ether in petroleum ether*)	*Fraction III* (*50% ethyl ether in petroleum ether*)
Aldrin	Chlorpham (CIPS)	Bensulide (Prefar)
Aroclors (PCBs)	Diazinon	Malathion[a] (Supracide)
Benefin	Dieldrin	Methidathion
BHC (α, β)	Endosulfan I (Thiodan I)	
CDEC (Vegedex)	Endrin	
Chlordane, technical	Malathion[a]	
Chlorpyrifos (Dursban)	Parathion	
DDE, DDD and DDT (*p,p*- and *o,p*-)	Parathion-methyl	
	SD 7438	
Ethion	Tetradifon (Tedion)	
Fonofos (Dyfonate)		
Guthion		
Heptachlor		
Heptachlor epoxide		
Isobenzan (Telodrin)		
Isodrin		
Lindane (γ-BHC)		
p,p-Methoxychlor		
Mirex		
Perthane		
Phorate (Thimet)		
Quentozene (PCNB)		
Ronnel		
Toxaphene		
Trifluralin		

[a] Malathion is an example of a pesticide that may give an inconsistent elution pattern requiring more than one eluting solvent.

the sample fractions quantitatively to graduated centrifuge tubes with 10–12 ml of 1% methanol in benzene or suitable solvent. Adjust the volume as necessary for the GC analysis, vortex to mix, stopper, and determine residues on GC.

The GC conditions for plant samples use the ^{63}Ni electron capture detector, which is equipped with a glass column 6 ft × $\frac{1}{4}$ inch i.d. packed with 5% OV-105, on Supelcoport 80/100 mesh packing. The column functions at its maximum with the column temperature at 205 °C, injection port 200 °C, electron capture detector 330 °C, and a carrier gas flow rate of 90 ml/min using nitrogen for the Tracor or 5% methane: 95% argon for the

Hewlett-Packard instruments. Any of the column packings given in Table 2 (p. 170) may also be used, with possible adjustments for optimum operating conditions.

Table 1 gives a few examples of pesticide elution patterns. There are many other pesticides not included here that would be partially eluted, not eluted at all, or would have inconsistent elution patterns using this particular solvent system.

5.2. Phenoxyalkanoic Acid Herbicides and Triazines

For phenoxyalkanoic acid herbicides such as 2,4-D and 2,4,5-T, weigh a 5 g chopped sample, transfer to an Omni-Mixer cup, add 50 ml of acidified acetone, pH 3 (4–5 ml of H_3PO_4 in acetone), and blend in the Omni-Mixer for 2 min. Filter the sample through a 'C' Buchner funnel into a 250 ml Erlenmeyer flask. Wash the Omni-Mixer cup with 2 rinsings of acidified acetone on to the funnel and wash the plant residue on to the filter with 100 ml of acidified acetone. Evaporate to 20 ml and take 10 ml, or an appropriate aliquot, and transfer to a 60 ml separatory funnel. Add 1 N KOH (approx. 10 ml) to pH 10 and mix. Wash the sample with 20 ml of ethyl ether, discarding the ether. Make the sample acid with 1–4 ml to pH 3. Extract sample with four 10 ml portions of ethyl ether, transferring the ether layer each time to a 150 ml beaker. Evaporate the ethyl ether just to dryness, and add 10 ml of BF_3 methanol. Heat on a water bath (95 °C) for 10 min and cool to room temperature. Transfer the sample to a 60 ml separatory funnel and rinse the beaker with 10 ml of distilled water. Add exactly 10 ml of hexane. Shake vigorously for 1 min, allow solvent layers to separate and drain off aqueous layer and save hexane (top layer). Wash hexane layer with 10 ml of distilled water 3 additional times, discarding the water each time. Transfer sample to a 10 ml graduated centrifuge tube and stopper. Adjust to desired volume and inject an appropriate size sample (1–10 μl) on the ^{63}Ni electron capture detector with a 6 ft glass column, $\frac{1}{4}$ inch i.d., packed with 1·5 % OV-17/1·95 % QF-1 on Supelcoport 100/120 mesh, with a column temperature of 180 °C, inlet 228 °C, detector 325 °C, and a 80 ml/min N_2 flow rate.

To determine triazines, weigh 10–50 g chopped sample, transfer to an Omni-Mixer cup, add 125 ml chloroform and grind at medium high speed for 4 min, or transfer sample to a 250 ml Erlenmeyer flask and shake for half an hour on a wrist action shaker.

Decant sample through a 22 mm i.d. column with 5 cm of anhydrous Na_2SO_4 into a 250 ml beaker, rinse flask with small portions of chloroform on to the Na_2SO_4 column, evaporate just to dryness on an airflow

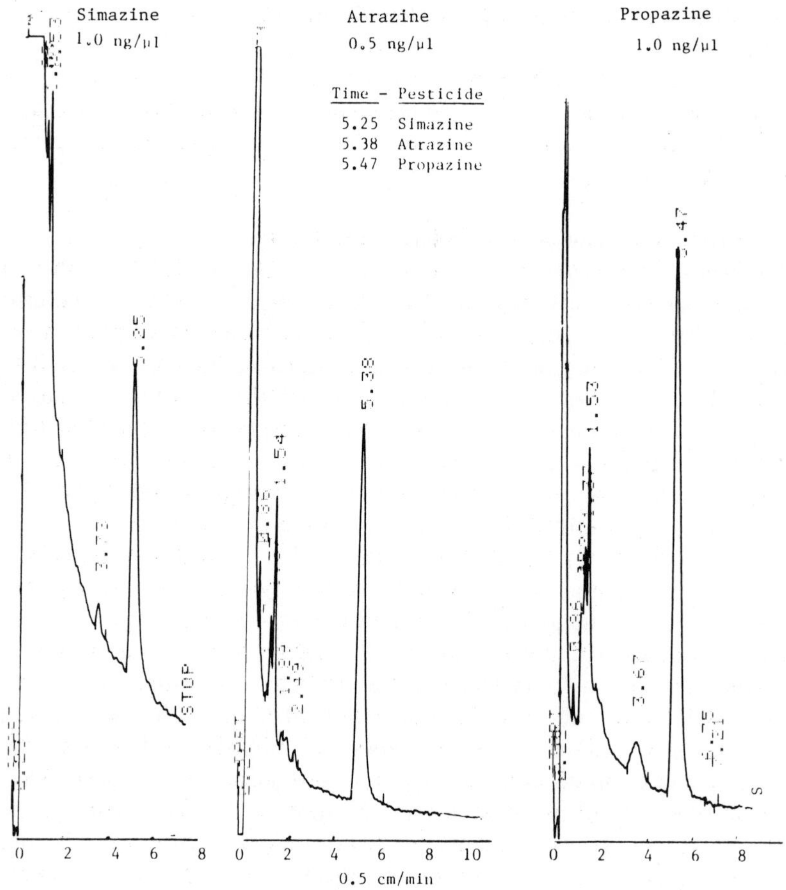

FIG. 1. Triazines.

Column:	3·6% OV-101/5·5% OV-210 on Chromosorb WAW DMCS
	80/100 mesh
	6 ft, $\frac{1}{4}$ inch i.d. glass
Detector:	electron capture ^{63}Ni
Temperatures:	detector 300 °C
	inlet 250 °C
	column 185 °C
Gas:	carrier 5% methane in 95% argon
	flow rate 45 ml/min

evaporator and take up immediately in 2 ml of benzene. Using a 15 mm i.d. column packed with 12 g activated Super I Woelm Basic Alumina (use 19 g of distilled water and 81 g of alumina, mix and place on a roller mixer for 1 h, let stand overnight before using), transfer the sample in benzene to the column with small rinsings of hexane, followed by a 75 ml hexane column wash, discard the hexane and elute the triazines from the column with 150 ml of 1:1 benzene–hexane mixture, collecting the sample in a 250 ml Erlenmeyer flask. Evaporate to dryness on a flash evaporator and if sample is oily, clean sample further by partitioning (5 times) as given in Section 5.1.1, substituting acetonitrile (saturated with hexane)–hexane for the acetonitrile (saturated with petroleum ether)–petroleum ether. At completion of partitioning, evaporate sample on a flash evaporator, just to dryness, and transfer quantitatively to a 10 ml graduated centrifuge tube with ethyl ether, evaporate the ether to 0·5 ml, make sample to desired volume with benzene, and inject on GC with a ^{63}Ni electron capture detector and a 6 ft glass column, $\frac{1}{4}$ inch i.d., packed with 3·6% OV-101/5·5% OV-210 on Chromosorb WAW DMCS 80/100 mesh, at a column temperature of 185 °C, injection temperature 250 °C and detector temperature 300 °C. Unless exact operating parameters are established, the three main triazines (simazine, atrazine and propazine) are extremely difficult to separate on the GC. An example of a successful separation is shown in Fig. 1. This triazine method is a modification of that of Mattson *et al.*[8]

6. MULTI-RESIDUE PROCEDURE FOR MEAT TISSUE AS RELATED TO FOOD PRODUCTS

A major food commodity consumed by the majority of the Western world's human population is meat, fresh or processed, such as poultry, fish, lamb, goat, beef or pork. To determine the pesticide contents of the above food products, a number of factors must be taken into consideration. A uniform procedure cannot be used for all meat tissue, as the tissue composition may vary between animal breeds, composition of food intake may have varied between animals, or the sample tissue may not be obtained from the same area of the animal.

Since pesticide residues tend to concentrate in fatty tissue, if possible, obtain the tissue from a fat layer. Or, if no fatty tissue is available, as in processed or lean meats, use what is available for the residue sample. As an example, when sampling fish, care must be taken always to take the tissue sample from the same section of the fish. If you wish to compare data of fish from various geographical areas, or from more than one fish in an area, sampling must be identical. Fat is concentrated in certain areas of a fish, and if the sample is taken from random areas of the fish, one may take the sample from a non-fatty area, thus obtaining a lower level of residue than if the sample had been taken from the fatty part of the fish.

6.1. Extraction of Meat or Fish

If fatty tissue is available from a meat sample, either put the fat on a filter paper in a glass funnel, stand in a glass beaker (250 ml), place in a forced draught oven at a suitable temperature (>90 °C), and heat until liquid fat drips into the beaker (half an hour or longer), or put the fat sample into a 150 ml glass beaker and place on a hot plate with a low heat setting, until a layer of liquid fat has been rendered from the fatty tissue. Weigh 1 g of extracted fat (or less, if 1 g is not available) and proceed to the partitioning step (Section 5.1.1).

If fatty tissue is not available, weigh a 25–50 g finely chopped tissue sample, place in a suitable size sharkskin filter paper (15 cm diameter), fold and staple closed. Weigh a 250 ml flat-bottom flask, containing 4 or 5 glass beads, accurately to the fourth decimal place, extract the sample in a Soxhlet extraction unit with petroleum ether (pesticide grade) for 6 h, drain any petroleum ether that may be present in the Soxhlet extraction tube into the lower flask, evaporate just to dryness on a rotary flash evaporator, cool and weigh the flask, determine the amount of fat in the tissue and save the extracted fat for the residue analysis. One must refer to the original weight of the fat in the meat sample to determine the concentration of pesticides in the sample. As above, weigh 1 g of extracted fat (or less, if 1 g is not available) and proceed to the partitioning step (Section 5.1.1).

If it is not convenient to extract the fat from the sample, one may choose to extract the residue from the fish or meat sample as outlined below. Weigh a 10 g sample of fish or meat and transfer to a 500 ml French square bottle, add 200 ml of 65 % acetonitrile in distilled water, grind for 30 s on a Polytron and filter through a 24 cm fluted filter paper (Schleicher & Schuell No. 588, fast flow), transfer 40 ml of the filtrate into a 250 ml separatory funnel, add 25 ml of 50 % petroleum ether in ethyl ether (pesticide grade). Shake for 1 min vigorously, venting as needed, with the stopcock to reduce the pressure, add approximately 10 g of sodium chloride (NaCl) plus 120 ml of distilled water, shake for 1 min as above, allow to stand for 10 min and drain aqueous layer into a second 250 ml separatory funnel. Add 25 ml of 50 % petroleum ether in ethyl ether to the aqueous layer in the second separatory funnel, shake vigorously for 1 min, venting as required, allow to separate for 10 min, drain off aqueous layer and discard. Combine the ether layers in the first separatory funnel, rinse the second funnel with petroleum ether and add to the first separatory funnel, add 25 ml of saturated NaCl solution, shake gently for 30 s, venting as required, allow 10 min for layers to separate, discard aqueous layer and pass the ether layer through a 4 cm column of anhydrous sodium sulphate (Na_2SO_4) (22 mm i.d.) into a 250 ml beaker to remove any water that may be present in the sample, follow by a 10 ml rinse of the funnel and Na_2SO_4 column with petroleum ether. If, after the sample has been evaporated *just to dryness*, a layer of fat or oil is seen on the sides of the beaker, the sample should be cleaned up using a partitioning and/or a column clean-up procedure.

6.2. Partitioning and Clean-up Procedure for Meat, Fish or Fat

If a noticeable fat layer from extracted meat or fish sample is found in the beaker, or if an extracted fat sample (1 g) is used, wash with 30 ml of petroleum ether into a 125 ml separatory funnel which already contains 30 ml of acetonitrile saturated with petroleum ether, stopper and shake on a mechanical shaker for 5 min, and allow 10 min for the layers to separate.

Drain the acetonitrile layer (bottom layer) into a 1000 ml separatory funnel and save. Similarly, extract the petroleum ether layer in the 125 ml separatory funnel 3 more times with 30 ml portions of acetonitrile saturated with petroleum ether, saving the acetonitrile layer each time as above. Discard the petroleum ether layer after the final shaking.

Add 100 ml petroleum ether to the combined acetonitrile extracts in the 1000 ml separatory funnel. Stopper and shake vigorously for 1 min, venting as needed with the stopcock to release any pressure.

Add 600 ml of distilled water and approximately 5 g of NaCl. Stopper, shake vigorously for 1 min, venting as necessary, and allow 10 min for separation of the solvent layers.

Drain the aqueous layer into a second 1000 ml separatory funnel. Add 100 ml of petroleum ether to the second separatory funnel containing the aqueous layer. Stopper and shake vigorously for 1 min, venting as necessary. Allow to stand and separate layers for 10 min.

Discard aqueous layer and combine the ether layers in the first separatory funnel. Use several small rinsings to clean the second funnel with petroleum ether and add the rinsings to the first separatory funnel. Add 10 ml of saturated NaCl solution and 100 ml of distilled water to the petroleum ether. Stopper and shake *gently for 30 s*, vent as necessary and allow 10 min for separation.

Discard the aqueous layer, and drain the ether layer into a 500 ml Erlenmeyer flask through a glass column (22 mm i.d.) of anhydrous Na_2SO_4 (4 cm), to remove any traces of water.

Evaporate *just to dryness* on a rotary flash evaporator and *transfer at once*, quantitatively, to a graduated centrifuge tube with hexane or appropriate solvent, and concentrate to 2 ml on an air or N_2 gas flow evaporator.

If the sample shows no indication of fat after partitioning, the next stage in the clean-up procedure for meats can be initiated in a number of ways. Two commonly used methods are given below.

6.3. Column Clean-up: Florisil

See Section 5.1.2.

6.4. Column Clean-up: Sep-pak Silica

Use the Sep-pak Silica cartridge (Waters Associates) by placing the elongated end of the cartridge in the lower tip of a 2 ml syringe barrel. Pre-wash the Sep-pak column with 2 ml of hexane and discard.

Place a clean graduated centrifuge tube under the Sep-pak. Add the sample (2 ml) from the partitioning step to the column and rinse the sample container with 2 small portions (1 or 2 ml) of hexane and add to the Sep-pak column.

Elute the Sep-pak column as follows. Add fraction I (6 ml of hexane) to the column. When the hexane has almost penetrated the top of the column, change to a clean graduated centrifuge tube and add the first portion of fraction II.

Fraction II is added as two separate portions, which prevents a sudden change in solvents. First add 4 ml of 3:1 benzene–hexane, and follow with 5 ml of 4:1 ethyl ether–benzene.

Evaporate each fraction to less than 0·5 ml and bring back to appropriate volume with 1 % methanol in benzene for injection on the GC. Stopper the centrifuge tubes and either run the samples at once, or store in an explosion-proof refrigerator until such time as the analyst is ready for the GC determination.

Refer to Section 5.1.2 for the column and operating conditions for the GC. The final determinative step in the analysis of pesticides is accomplished on the GC units using the various detectors to determine and confirm the residue data.

7. ANALYSIS OF WATER FOR PESTICIDES

Water is an important material used in conjunction with food preparation and may be contaminated with many water soluble pesticides. To ensure that sufficient sample is available for the analysis of the major groups of pesticides often found in water, a gallcn of water should be obtained in a chemically clean glass bottle.

If the water is dirty, it should be filtered through No. 2 Whatman filter paper before proceeding with the extraction. To determine chlorinated hydrocarbons, measure an 800 ml water sample, transfer to a 1 litre Erlenmeyer flask, add 100 ml of solvent (90 % benzene in hexane), and stir with a magnetic stirrer for 1 h. Transfer sample into a 1000 ml separatory funnel and allow to stand for 15 min. Draw off aqueous layer (bottom layer) into the 1000 ml Erlenmeyer flask, drain the benzene–hexane layer through a 22 mm i.d. column of anhydrous Na_2SO_4 (4 cm) into a 500 ml Erlenmeyer flask and add a second 100 ml volume of solvent (90 % benzene in hexane) to the 1000 ml Erlenmeyer flask and repeat stirring as above for a second hour, discarding the aqueous layer after the second extraction. Evaporate the benzene–hexane layer on a rotary evaporator. *Do not allow the sample to go to dryness.* Transfer to a 15 ml centrifuge tube with 1 % methanol–benzene. Adjust the volume to a suitable concentration and inject on GC.

To determine the presence of the phenoxyalkanoic acid herbicides (2,4-D, 2,4,5-T, Silvex, etc.), measure 500 ml of water and transfer to a 1000 ml separatory funnel (with Teflon stopcock). Tilting the separatory funnel carefully, sprinkle several spatula scoops of Na_2SO_4 over the water. Stopper and shake vigorously. Repeat as necessary until water is saturated. Acidify to a pH of 2 (with pH paper) using approximately 1–5 ml of phosphoric acid (H_3PO_4). Add 100 ml of ethyl ether and shake the separatory funnel vigorously for 1 min, releasing the pressure by venting with the stopcock. Allow the solvent layers to separate for at least 10 min. Drain off the aqueous layer into a second separatory funnel, and drain the ethyl ether into a 250 ml Erlenmeyer flask. Extract the water a second time with 50 ml of ethyl ether as above, allowing separation of solvent layers, discard the aqueous layer (bottom layer), and combine the ether layers in the 250 ml Erlenmeyer flask. Evaporate the ether layer to about 5 ml on a flash evaporator and transfer to a 50 ml Teflon-lined screw top test tube, rinse out the Erlenmeyer flask with small portions of ethyl ether and add to the 50 ml tube. Evaporate the sample to near dryness on a N-Evap evaporator (Organomation Associates, Inc., Shrewsburg, MA) with air or nitrogen flow and continue with methylation.

Methylation may be accomplished by using boron trifluoride in methanol (BF_3 methanol) or diazomethane.

7.1. Methylation: BF_3 Methanol

Add 10 ml of BF_3 methanol and heat to near boiling (85–90 °C) in a water bath for 15 min to esterify, then cool to room temperature. Transfer the sample to a 60 ml separatory funnel and rinse tube with 15 ml distilled water into the separatory funnel, followed by rinsing the tube with 10 ml hexane and adding this also to the separatory funnel. Shake the separatory funnel vigorously for 1 min and allow layers to separate for 10 min. Drain off the aqueous layer (lower layer) and wash the sample two more times with distilled water, discarding the aqueous layer (lower layer) each time. Transfer the hexane layer through a column (22 mm i.d.) of anhydrous Na_2SO_4 (4 cm) into a 10 ml graduated centrifuge tube, adjust the volume for GC analysis and inject on a GC unit equipped with a ^{63}Ni electron capture detector.

7.2. Methylation: Diazomethane

When using diazomethane, greater care must be taken owing to the explosive and poisonous nature of this chemical. It requires the use of a *high draught hood*, gloves and goggles. Add 0·5 ml isooctane to the sample, then add diazomethane dropwise with agitation after each addition, until a definite yellow colour persists (1–5 ml). Let stand for 15 min. Bubble nitrogen through the solution until the yellow colour disappears (5–10 min). Continue evaporation to about 0·3 ml. Transfer sample quantitatively to a centrifuge tube. Dilute or concentrate to desired volume with benzene and inject on the GC.

To prepare diazomethane, in an Erlenmeyer flask that has not been etched and *without* a ground glass joint, dissolve 2·3 g potassium hydroxide (KOH) in 2·3 ml distilled water. Let cool to room temperature and add 25 ml hexane, swirling to mix. Cool flask in freezer for 15 min. In a *high draught hood* and using gloves and goggles, add 1·6 g nitrosomethyl urea, very slowly, mixing the contents of the flask after each addition. (There will be some foaming.) Decant the hexane into a tube with a Teflon-lined screw top. This may be stored for one week at −18 °C. Diazomethane methylation is adapted from the *Manual of Analytical Methods for the Analysis of Pesticides in Human and Environmental Samples*[9] and private correspondence with Consolidated State Laboratory, Richmond, VA.

To determine the presence of triazines, measure a 400 ml water sample, transfer to a 1000 ml Erlenmeyer flask, add 150 ml of chloroform and 5–10 mg sodium chloride. With a magnetic stirrer, stir for 20 min. Pour sample into a 1000 ml separatory funnel and let layers separate for about 20 min. Drain chloroform (lower layer) through a 22 mm i.d. column with 4 cm of anhydrous sodium sulphate into a 600 ml beaker. Repeat this extraction two more times with 100 ml of chloroform each time. Evaporate the chloroform on an airflow evaporator almost to dryness. Add approximately 20 ml of benzene and repeat evaporation almost to dryness. Transfer sample quantitatively to a graduated centrifuge tube with 10–12 ml of benzene. Adjust volume as necessary for injection of sample into the GC unit.

8. OPERATIONAL GUIDE FOR GAS CHROMATOGRAPHY

Gas chromatography[10] is a technique for separating volatile substances by percolating a gas stream over a stationary phase, which is either a solid (GSC) or a liquid (GLC) spread as a thin film over an inert solid. The basis for separation in GLC is partitioning of the sample in and out of this liquid film. Samples are identified qualitatively by retention time (the time required for the sample to pass through the column from injection of sample to its detection), whilst quantitative data are obtained by measurement of the area produced under each peak and relating this area to the concentration of the standard. It is important that no leaks exist anywhere in the flow system. At the injection port a leak may cause the sample to disperse into the atmosphere before it can enter the column to be detected. Slow leaks may adversely affect peak retention times. Standard solutions of the known compounds are run on the GC and the unknown compound is compared to the known. Glass columns should always be used for a pesticide column as many of the pesticides may react with copper or stainless steel, resulting in degradation of the pesticide. To ensure that column packings are standardised, they should be purchased from a chemical supply house such as Supelco, Alltech Associates, etc. The glass columns may be packed in the laboratory by placing silanised glass wool into one end of the column, connecting suction on the plugged column end, and using a funnel on the other end of the column to pour the packing into the column. Either gently vibrate or gently tap the column until it is filled. Put silanised glass wool into the open end of the column and place column in GC oven. Condition with a N_2 stream blowing through the column with the oven at a temperature 25 °C higher than operating temperatures, unless this exceeds the maximum temperature for the liquid phase of the packing. The column should not be connected to the detector until it has been conditioned.

Commonly used carrier gases are helium, nitrogen, and 95 % argon/5 % methane. Hydrogen, oxygen and breathing air are used for the hydrogen flame ionisation detector. All gases must be of a high-purity grade or the column may be inadvertently damaged, the electron capture detector may be harmed or unwanted peaks may be produced.

In the initial installation of the instrument, the column temperature should not be turned on until the detector has heated up to temperature, and the temperature of the detector should always be higher than that of the column to avoid collection of low-temperature products on the detector, making it insensitive. The injection port should always be operated at a temperature higher than the column temperature to avoid condensation of the sample.

The sample is injected, usually with a 10 μl syringe, through a septum that is placed in the injection port above the column to help prevent leaks. Excellent results have been obtained using silicone septa. Care should be taken to ensure that the septum does not develop a leak, at which point it should be changed at once. A septum should be changed at the end of the working day so that the septum is equilibrated for the next day's work period.

In connecting the GC columns, either an 'O'-ring (a heat-resistant silicone rubber ferrule) or a Teflon front ferrule should be used with a brass back ferrule. Make sure that all connections are tightened firmly, but not over-tightened, to prevent breaking the glass column.

9. GENERAL OPERATING CONDITIONS FOR GC

The standard operating conditions for detection of multiple residue pesticides may vary owing to column type, instrument used and pesticide(s) under study. For example, the standard operating conditions for general chlorinated hydrocarbons on column packing OV-105 are as follows:

Column packing:	5 % OV-105 on Supelcoport 80/100 mesh
	6 ft, $\frac{1}{4}$ inch i.d. glass
Detector:	electron capture ^{63}Ni
Temperatures:	detector 330 °C
	injection port 220 °C
	column 205 °C

Gas: carrier—nitrogen
Rotameter 9·0
40 lb/in^2
flow rate 90 ml/min

Typical chromatograms for organochlorine pesticides are shown in Figs 2, 3 and 4. If all these residues were present in a single sample, the problem of peak identification would be difficult. If one were to superimpose Figs. 2, 3 and 4, one could see how it would be extremely

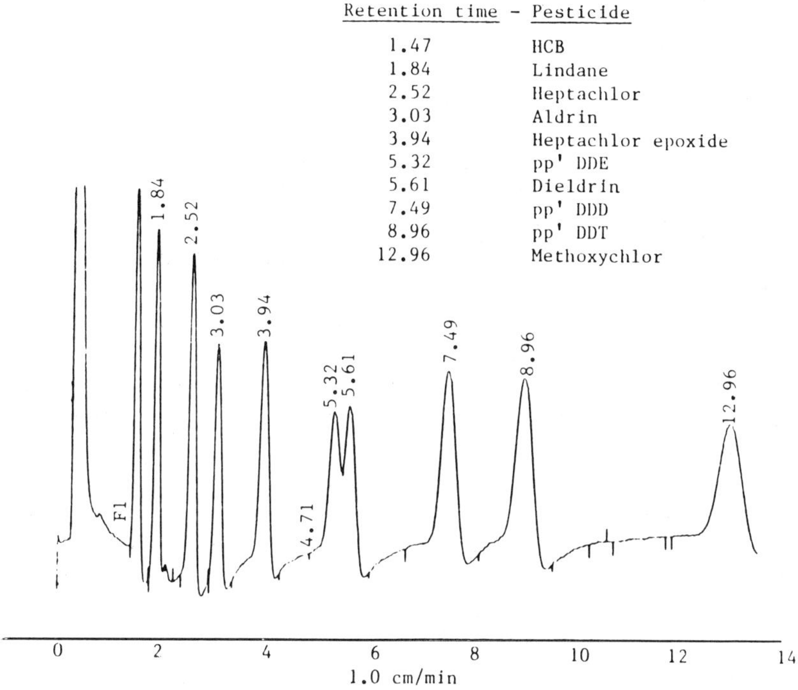

FIG. 2. Organochlorine pesticides.

Column: 5% OV-105 on Supelcoport 80/100 mesh
6 ft, $\frac{1}{4}$ inch i.d. glass
Detector: electron capture ^{63}Ni
Temperatures: detector 330 °C
inlet 220 °C
column 205 °C
Gas: carrier N_2
Rotameter 9
40 lb/in^2, 90 ml/min

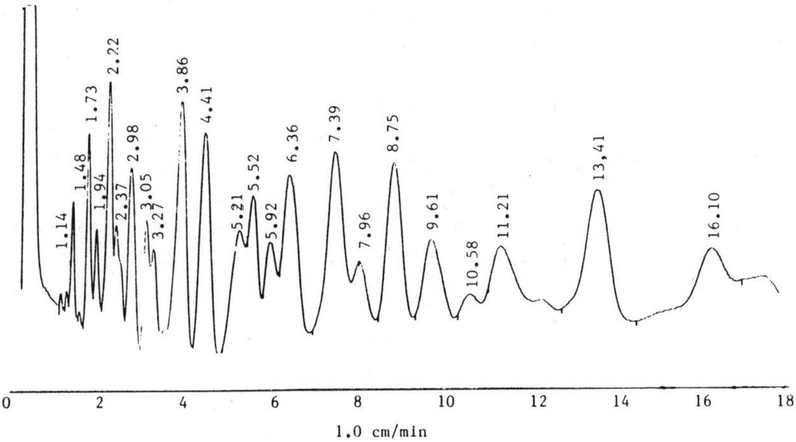

FIG. 3. Polychlorinated biphenyls (mixture of PCB 1242, 1254 and 1260, 0·5 ng/μl). For operating conditions see legend to Fig. 2.

difficult to differentiate the Aroclors, toxaphene, chlordane and the various organochlorine pesticides individually listed in Fig. 2.

Since the electron capture detector will detect most volatile organic compounds, it is often necessary to establish verification of a residue peak. There are often extraneous peaks and overlapping peaks in residue samples. This necessitates using a variety of means to attempt peak confirmation. The use of pairs of working columns with dissimilar polarities can be useful in this identification. Compounds which are detected on one column packing may not be detected on another dissimilar column packing, whereas the standard of the suspected compound would be detected on both columns. Also, if the compound is eluted in a different sequence on the two dissimilar columns, the sample peak should also be eluted in the same manner on both columns or it cannot be established that the questionable peak is the suspected compound. An example of complementary columns for pesticide analysis is OV-17/QF-1 and OV-210.

Verification of residues containing phosphorus or sulphur may be accomplished by using the flame photometric detector in the appropriate mode. The two modes of the flame photometric detector are phosphorus (526 μm filter) and sulphur (394 μm filter). Each of these modes is specific and will detect *only* those compounds containing phosphorus or sulphur. It should be noted that the sulphur mode is an order of magnitude less sensitive than the phosphorus mode. Examples of phosphorus compounds, detected by both ^{63}Ni electron capture and flame photometric detectors,

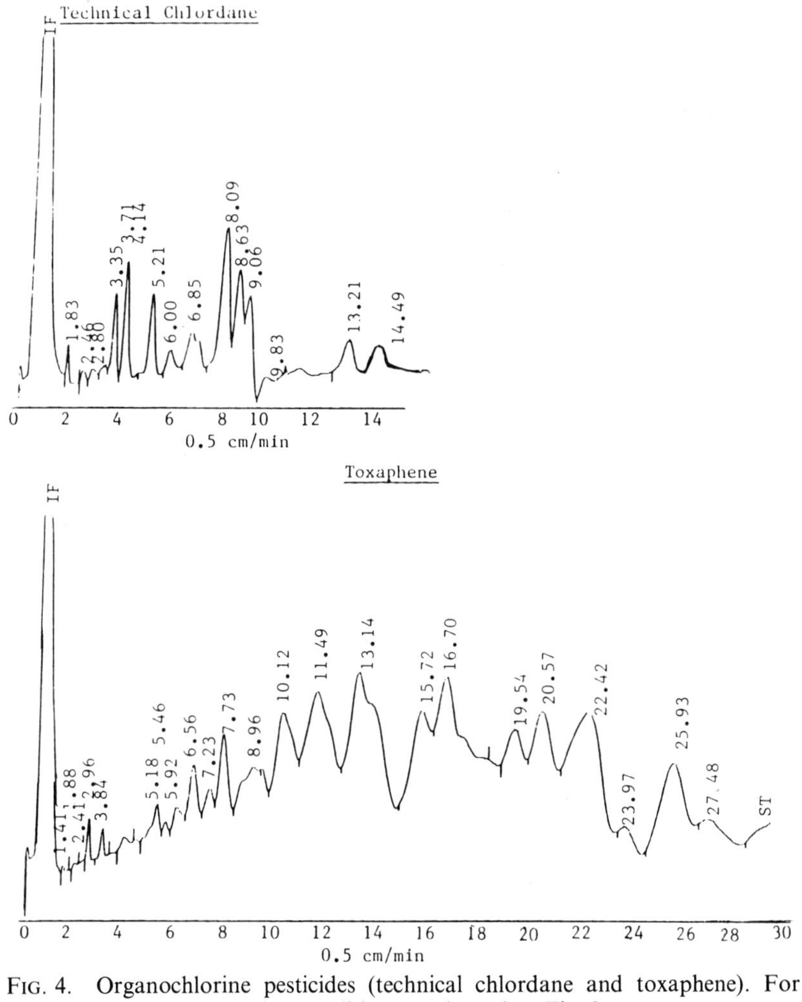

FIG. 4. Organochlorine pesticides (technical chlordane and toxaphene). For operating conditions see legend to Fig. 2.

are shown in Figs. 5 and 6. Other available specific detectors include the N–P detector for nitrogen and phosphorus and the Hall detector for chlorine and nitrogen compounds.

As a further means of identification, thin-layer chromatography techniques, chemical derivatisation of the compound in question, high-pressure liquid chromatography and mass spectrophotometry are accepted as additional means of positive identification of questionable compounds.

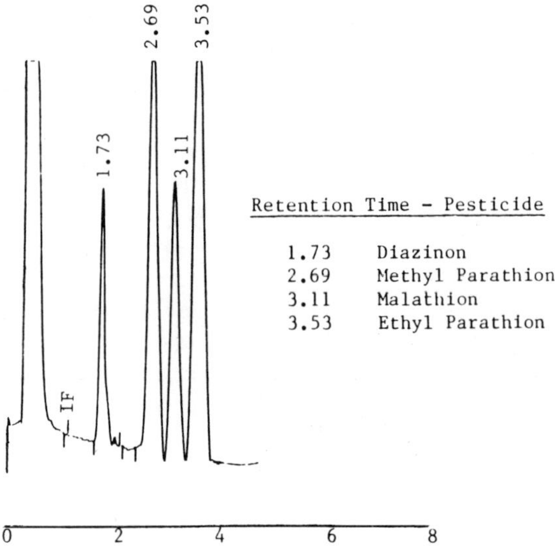

FIG. 5. Organophosphorus pesticides: electron capture detector. For operating conditions see legend to Fig. 2.

There are a wide variety of column packing materials available which are suitable for pesticide analysis. Some of the packings currently in use are shown in Table 2. To select a column, one should choose a packing that is capable of separating a large number of pesticides with a minimum of overlapping. The column should also be of high efficiency to detect low pesticide concentrations, with low bleed to avoid extraneous peaks and with high maximum temperature to provide a more stable column with less column bleed.

TABLE 2
GC COLUMNS USED IN PESTICIDE ANALYSIS

Support	*Stationary phase*
Chromosorb W HP 100/120 mesh	1·5% OV-17 and 1·95% QF-1
Chromosorb W HP 100/120 mesh	1·5% OV-17 and 1·95% OV-210
Supelcoport 100/120 mesh	1·5% SP-2250 and 1·95% SP-2401
Supelcoport 80/100 mesh	3% SP-2100
Supelcoport 80/100 mesh	5% SE-30
Chromosorb WAW DMCS 80/100 mesh	3·6% OV-101 and 5·5% OV-210

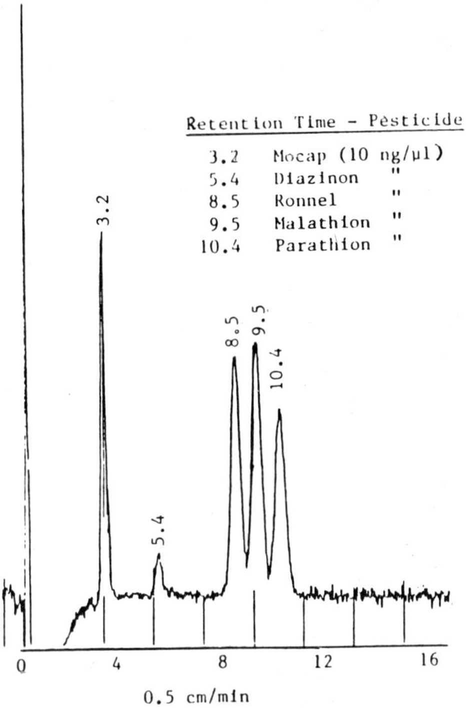

FIG. 6. Organophosphorus pesticides: flame photometric detector.

Column:	10% SP-2100 on Supelcoport 80/100 mesh 6 ft, $\frac{1}{4}$ inch i.d. glass
Detector:	flame photometric phosphorus mode, 526 μm
Temperatures:	detector 225 °C inlet 220 °C column 205 °C
Gas:	carrier N_2 Rotameter 9 PSI 40–90 ml/min
Detector gases:	H_2—Rotameter 70 20 lb/in^2, 35 ml/min Air—Rotameter 70 80 lb/in^2, 150 ml/min O_2—Rotameter 2 80 lb/in^2, 150 ml/min

Although it is important that the correct column packing be selected, many other factors should be considered. The best results are obtained using glass columns, usually 6 ft in length, with an i.d. of $\frac{1}{4}$ inch or $\frac{1}{8}$ inch, depending upon the elution speed desired. The $\frac{1}{8}$ inch i.d. gives a faster elution pattern than the $\frac{1}{4}$ inch i.d. column. Also, columns of less than 6 ft may be adequate when working with late-eluting compounds like mirex and methoxychlor. For scanning samples with multiple residues, a 6 ft column is recommended to obtain maximum column efficiency and peak resolution.

Because the analysis of pesticides is often concerned with low concentrations (ppm, ppb), it is imperative that a detector with high sensitivity be employed. Factors affecting the sensitivity of the GC detector include carrier gas flow rate, cleanliness of the sample, signal-to-noise ratio of the instrument, injection port, column oven and detector temperatures, concentration of the sample and the amount of sample injected.

Care should be taken to work within the linear range of the detector. In order for the detector to be operated at maximum efficiency, the linearity range of each compound of interest should be established.

Proper adjustment of the column temperature and carrier gas flow rate is essential to good column efficiency, proper resolution of peaks, proper peak height response, proper elution time and stable baseline performance. Often only through trial and error, one's understanding of one's own instruments and the persistence of the analyst, will maximum efficiency and the best combination of operating parameters be maintained.

While there are several means of quantifying chromatographic peaks, it is important that some uniformity in procuring final data be established. Interpretation of the chromatogram, choice of using peak area or peak height as a measure of amount of compound present, and the incorporation of all factors to the analysis which may affect the outcome of the final calculation are necessary considerations of the analyst. In pesticide residue analysis where there are often overlapping peaks on the chromatograms, concern must be given to side-arm peaks, peaks on a slope and double peaks in addition to peaks having a straight baseline. There are several methods of calculation available. The only one we shall be concerned with is the external standard method. In this method, pesticide concentration is determined by direct comparison of the peak of the sample to the standard. Care should be taken to keep the injection volume and response of the standard as close as possible to that of the sample. Samples may be concentrated or diluted as necessary to keep detector response to sample and standard within an acceptable peak height variance of 10–25%.

The results of the data collected from a GC chromatogram are calculated using the following universal equation:

$$\mathrm{PR} = \frac{P_s H_p D}{P_s W}$$

where PR = concentration of pesticide residue in sample, P_s = ng of pesticide standard, H_p = peak height or area of sample, D = dilution factor, P_s = peak height or area of standard and W = weight of original sample (in g or ml). To determine the value of the dilution factor (D),

$$D = \frac{\text{Initial extraction volume (ml)} \times \text{Final volume (ml) before GC injection}}{\text{Total or aliquot of initial extraction volume (ml)} \times \mu\text{l injected on GC}}$$

Example

A 25 g sample of apples was extracted with 200 ml of solvent, a 50 ml aliquot was taken, the final volume of sample before injection on to GC was 5 ml and the amount injected was 2 μl. The concentration of the endrin standard was 0·05 ng/μl and 2 μl of standard was injected for a peak height of 110 mm. The 2 μl sample injection gave a peak height of 92 mm.

Utilising these data:

$$\mathrm{PR} = \frac{0{\cdot}1 \times 92 \times D}{110 \times 25}$$

where

$$D = \frac{200 \times 5}{50 \times 2} = 10$$

therefore

$$\mathrm{PR} = \frac{0{\cdot}1 \times 92 \times 10}{110 \times 25} = 0{\cdot}0335 \text{ ppm endrin.}$$

ACKNOWLEDGEMENTS

The author wishes to thank Jean Dickinson for her technical assistance and Joyce Burroughs for reading the manuscript.

REFERENCES

1. THOMPSON, J. F. (ed.) (1976). *Manual of Analytical Quality Control for Pesticides in Human and Environmental Media*, Sect. 7, EPA-600/1-76-017 (Feb.).
2. WATTS, R. R. (ed.) (1980). *Manual of Analytical Methods for the Analysis of Pesticides in Human and Environmental Samples*, Sect. 5, EPA-600/8-80-038 (June).
3. BURKE, J. A. (1965). *J. Ass. Off. Anal. Chem.*, **48**, 1037.
4. (a) McMAHON, B. M. and SAWYER, L. D. (eds.) (1968). *The Pesticide Analytical Manual*, Vol. 1, US Food and Drug Administration, Washington, DC; revised Sept. 1982. (b) MARCOTTE, A. L. and BRADLEY, M. (eds.) (1971). Ibid., Vol. 2; revised Dec. 1982.
5. MILLS, P. A. (1968). *J. Ass. Off. Anal. Chem.*, **51**, 29.
6. ARMOUR, J. and BURKE, J. (1970). *J. Ass. Off. Anal. Chem.*, **53**, 761.
7. *Pesticide Analytical Manual*, Vol. 1, Sect. 251, 12a (Achnion).
8. MATTSON, A. M., KAHRS, R. S. and SCHNELLER, J. (1965). *J. Agric. Fd Chem.*, **12**, 120.
9. Ref. 2, Sect. 5(A)(4)(c), p. 3.
10. McNAIR, H. M. and BONELLI, E. J. (1969). *Basic Gas Chromatography*, 5th Edn, Varian, Inc., Walnut Creek, CA, pp. 1–6.

Chapter 5

IMMUNOCHEMICAL METHODS IN FOOD ANALYSIS

JEAN DAUSSANT and DANIELLE BUREAU

CNRS, Laboratoire de Physiologie des Organes Végétaux, Meudon, France

1. INTRODUCTION

The specificity of the antigen/antibody reaction and the high sensitivity of immunochemical methods provide a unique means for detecting and determining the amount of one constituent in a complex mixture, even if the constituent is present in minute amounts. These immunochemical techniques play an important part in clinical analysis and constant progress in their development is opening new fields of application.

The methods began to be applied to food analysis about two decades ago. However, routine applications on a large scale are still hampered by a few specific difficulties. When immunochemical methods are carried out on fluids which have not been subjected to heat treatment, such as raw milk, the analysis is the same as in clinical laboratories. These immunochemical methods can be applied in a similar way whatever the nature of the constituents, be they macromolecules or small molecules. However, the problem is generally complicated by at least two factors: in many cases the constituents have to be determined in solid products rather than fluids, and their exhaustive extraction, without denaturation, constitutes an additional and sometimes difficult step. The second and major factor is that the products to be analysed (solids or liquids) may have been submitted to technological processes involving physico-chemical and thermic procedures during their production and sterilisation. Thus, the identification and quantification may concern denatured constituents which have lost part or all of their original antigenic specificity and solubility. Therefore, immunochemical methods when applied to foods encounter specific difficulties which do not exist in clinical analysis.

The many studies on immunochemical techniques in food analysis have been reviewed in recent papers.[1–9] Therefore, this chapter aims at evaluating the present situation of the application of different immunochemical techniques in food analysis. Moreover, it aims at focusing on some particular characteristics, problems and trends of the immunochemical approach in food analysis.

2. ANTIGENS, ANTIBODIES, PRODUCTION OF ANTIBODIES

Before using immunochemical methods for the detection and quantification of one constituent (antigen), the biological reagents involved in these methods (antibodies) have to be prepared in higher vertebrates by immunising them with the antigen. The preparation of the biological reagents is based on one of the characteristics of the defence system of higher vertebrates which aims at recognising and then rejecting any foreign constituent invading their organism. The formation of antibodies, in response to the injection of one antigen, is one of the expressions, at the molecular level, of the defence system of the higher vertebrates.

2.1. Antigens

Antigens are characterised by their antigenicity and their immunogenicity. The immunogenicity resides in their capacity of inducing the formation of antibodies. The immunogenicity depends to some extent on the size of the antigen. Small molecules, like hormones, are not immunogenic; they are called 'haptens'. Once the hapten is bound to a large carrier, generally a protein, it becomes immunogenic. The antigenicity, or antigenic specificity, characterises the antigen capacity of reacting with the antibodies. The reaction involves small sites on the surface of the antigenic molecules called antigenic determinants, or epitopes. In proteins, the antigenic determinants are formed by a reduced number of amino acids. A distinction has been made between 'sequential' and 'conformational' determinants.[10,11] Recently, the distinction was defined more precisely and the terms 'continuous' and 'discontinuous' were proposed for the determinants.[12] According to this definition, the continuous determinants are built up with amino acids which are contiguous in the primary structure, whereas the discontinuous determinants are made with amino acids which are not contiguous in the primary structure. Immunisation with native proteins mainly induces the formation of antibodies specific for discontinuous determinants, whereas the use of denatured proteins favours the induction of antibodies specific for continuous determinants.[13] Antigens such as bacteria, cells, viruses or proteins each have several different antigenic determinants some of which may be present at different places in the molecule (repetitive determinants), as is the case for proteins made of identical subunits. Therefore these antigens are multivalent.

2.2. Antibodies

Antibodies belong to the family of the immunoglobulins which represent about 20 % of the serum proteins. The immunoglobulins are formed by two identical heavy chains and two identical light chains (Fig. 1). Disulphide bridges bind the two heavy chains together; they also bind the light chains, each to one distinct heavy chain. There are five classes of immunoglobulins, called IgA, IgG, IgM, IgD and IgE, which differ from each other by the type of heavy chains. IgA and IgM exist respectively in dimeric and pentameric forms. IgA, IgG and IgM are involved in the immunochemical methods to be described, but IgG, which represents about 75 % of the total immunoglobulin, is the main immunoglobulin involved. Each monomeric form of one antibody molecule has two antibody sites, also called paratopes. They are situated between the light and the heavy chains towards the N-terminal end of these chains. Each of the two paratopes of

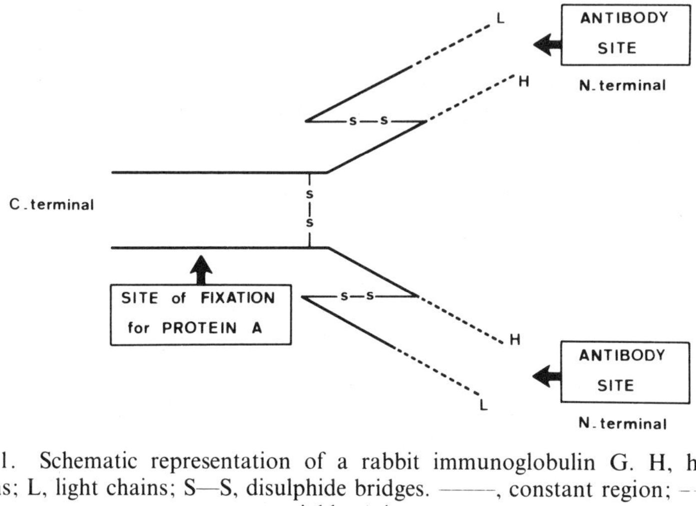

FIG. 1. Schematic representation of a rabbit immunoglobulin G. H, heavy chains; L, light chains; S—S, disulphide bridges. ——, constant region; ----, variable region.

one antibody molecule is specific for the same epitope. Thus, one monomeric antibody molecule may react with two identical epitopes located on the same or on two distinct antigen molecules. The reaction between antigens and antibodies involves non-covalent bonding, the strength of which is in relationship to the complementarity of the corresponding epitope and paratope structures.

One further characteristic of antibodies is worthy of mention because it is involved in immunochemical techniques used in food analysis. It concerns the capacity of protein A from *Staphylococcus aureus* to bind specifically to the IgG molecules of many animal species (with the exception of one human IgG subclass) without any interference to the antigen/antibody reaction.[14] The characteristics related to structural features of the IgG molecules are shown schematically in Fig. 1. The immunological background concerning the structural and functional aspects of antigens, antibodies and their interaction can be found in manuals or general reviews such as refs. 15–17.

2.3. Production of Antibodies

Antibodies are produced by immunising higher vertebrates with the purified antigen. Rabbits, mice, guinea pigs, goats, sheep, hens and horses are used. Several factors have to be considered for the immunisation:

1. The route of immunisation can be intravenous, intramuscular, subcutaneous or intraperitoneal.
2. Adjuvants, which facilitate the production of antibodies. There are two sorts of adjuvants, the complete and the incomplete. The incomplete adjuvant acts as a depot of antigen in the organism: the antigen is injected in emulsion with mineral oil. In addition, the complete adjuvant stimulates the defence system of the animal: killed mycobacteria are injected together with the antigen in the emulsion.
3. The frequency.
4. The number of injections.
5. The amount of antigen injected; this depends on the immunogenicity of the antigen and on the two preceding factors.

The serum collected after immunisation is called immune serum. The serum collected before immunisation is called pre-immune or null serum. The latter is used in order to ascertain that the reactions obtained with the immune serum are due to the antigen/antibody reactions. An immune serum specific for one constituent is called monospecific immune serum.

An immune serum containing several sets of antibodies each specific for different antigens is called polyspecific immune serum. There is no general procedure for immunisation. Details and examples of immunisation procedures have been reported.[15,18–21]

In the case of haptens, which are molecules too small to be immunogenic, the immunisation is carried out as for other antigens but using the hapten previously covalently bound to a carrier, generally a protein. The number of haptens bound to the carrier and the eventual damaging effect resulting from the binding may be important for the quality of the immune serum obtained.

The immune sera can be used as such in a number of techniques. However, in certain techniques it is better to separate the IgG fraction from the other serum proteins. This is achieved by using ammonium sulphate precipitation or by combining this procedure with chromatography on DEAE cellulose.[22,23] Alternatively, protein A immobilised on a support can be used in affinity chromatography for separating the IgG.[24]

In addition to these traditional ways of preparing antibodies, a procedure for immunising hens and then using the antibodies of the egg yolk is worthy of mention. This procedure does not involve bleeding the animals, and immunoglobulins are easily separated from egg yolk, so a considerable amount of antibody can be obtained in a relatively short time.[21]

The progress recently made in the production of monoclonal antibodies[25] offers the possibility of obtaining new biological reagents. The antibodies are produced by somatic cell hybrids (myeloma cells and spleen cells from immunised mice) grown in tissue culture. Large amounts of antibody specific for a single epitope can be obtained in this way.

3. IMMUNOCHEMICAL METHODS USED IN FOOD ANALYSIS

The principles and applications of immunochemical techniques can be found in numerous textbooks or review articles (refs. 26 and 27 for example). This section deals with some particular points concerning the application of immunochemical methods employed in food analysis. The principle of these techniques will be briefly recalled.

3.1. Passive Agglutination Techniques[28,29]

When the antigen is particulate, the action of antibody on the antigen results in visible agglutination. An agglutination reaction can also be carried out with non-particulate antigens such as polysaccharides, proteins or haptens, if red cells or latex particles are coated (sensitised) with the antigen or the hapten. The action of the anti-antigen or anti-hapten antibodies on the sensitised particles results in their agglutination. This is the so-called passive agglutination technique. Standard conditions for the reaction between the sensitised particles with serial dilutions of the immune serum must be established. The highest dilution of the immune serum for which the agglutination is still visible is referred to as the titre of the immune serum. Starting from these conditions, the immune serum and free antigens are first incubated together and then added to the sensitised cells. Some of the antibodies react with the free antigens and are therefore not available for the reaction with the sensitised cells, and the agglutination becomes inhibited. There is a relationship between the extent of the inhibition and the amount of free antigen preincubated with the immune serum.

These techniques are well adapted to food analysis, particularly in serial investigations, provided that a sufficient number of parallel test series are carried out. However, many of the results obtained have given rise to controversy (see ref. 4 for review). Therefore these techniques are no longer used much today, probably because of the considerable progress made in other methods.

3.2. Precipitation Techniques

When soluble macromolecular antigens are incubated together with their corresponding antibodies, immunecomplexes form and precipitate under certain conditions. The precipitation probably results from the loss of polar groups on the immunecomplexes. The precipitation depends on many factors such as the class of immunoglobulins and the nature of the antigen, the concentration of the reactants, ionic strength, pH and temperature.[30] The precipitation reaction is the basis of a considerable number of techniques.

3.2.1. Ring Test Technique

This is a classical technique which is cheap and rapid for detecting one protein in a mixture. It is now occasionally used in food analysis.[31] Tubes of about 2 mm diameter and a few cm in height are half filled with the immune serum or with its dilution. Serial dilutions of the antigen solution are added on top of the immune serum. The formation of a precipitate at the interface of the immune serum and the antigenic solution is noted after different periods of time ranging from 15 min to several hours.

3.2.2. Quantitative Precipitation Determination

A quantitative evaluation of the precipitate can be carried out after its centrifugal separation from the supernatant by using a protein determination.[32] When increasing amounts of antigen are added to the same amount of the corresponding immune serum, the quantity of the precipitate increases, reaches its maximum at a certain ratio between antigen and antibody concentrations and then decreases. When there is an excess of antigen, part of the immunecomplex remains soluble. A reference scale can be established between the amount of antigen added and the quantity of immunoprecipitate. The ascending part of the precipitation curve, excess antibody, is used for establishing the reference scale. This classical technique has been used frequently for quantifying one antigen in a protein mixture, particularly for the determination of foreign proteins in meat products (see ref. 4 for review). However, the estimation of antigen based on the measurement of immunoprecipitates separated from the initial antigen/antibody mixture is time-consuming and not suitable for routine analyses. In contrast, the measurement of the immunecomplex during its formation provides a means of rapid determination by using nephelometric and turbidimetric techniques.

3.2.3. Nephelometric and Turbidimetric Techniques

The estimation of the amount of precipitate can be achieved either by measuring the scattered light (nephelometry) or by determining the transmitted light which is not scattered by the particles in suspension (turbidimetry). These techniques need careful standardisation.[33]

To measure scattered light requires the use of a nephelometer or photronreflectometer.[34] This method has been used for examining the level of IgG in a great number of milk samples in order to permit the detection of colostrum.[35] The method was used manually: a series of reference tests as well as the unknown samples (six dilutions of each) were determined under the same conditions (same amount of immune serum and temperature of incubation). Measurements were made at 632·8 nm after 30 min incubation. The method is rapid and it is possible to take one reading every 15 s; also, each determination involved the use of only 40 μl of immune serum.[36]

Turbidimetric analysis requires no specific apparatus; spectrophotometers or even colorimeters may do. The technique has been used in clinical analysis for the determination of immunoglobulins and in combination with the use of a centrifugal analyser it can be speeded up and automated.[37,38] The centrifugal analysers which have revolutionised the automation of food analysis, especially with colorimetric and enzymatic methods, are now also being used for immunoturbidimetric methods.[39] The immune serum is diluted in a solution of polyethylene glycol for accelerating the formation of the precipitating immunecomplex. The different dilutions of the antigen samples (unknown samples as well as reference samples) are automatically pipetted in separated compartments of a cell. The centrifugation brings the reactants together at the same time, and spectrophotometric determinations (290 nm) are carried out during the centrifugation at 10 s intervals. The evolution of the extinction difference by units of time is reported only in the area corresponding to a constant increase in the extinction. The reference scale is plotted by giving the mg of antigen against the extinction difference by unit of time. The method delivers very reproducible results due to the highly automated device; it takes about 10 min for the analysis of five samples (including their dilution and the reference scale) and consumes, depending on the system, 0·5–2 ml immune serum (average 30 μl by determination).[40] The method has been commonly used in the last few years for determining several specific proteins in different food products.[40]

3.2.4. Precipitation Techniques Carried out in Gels

A considerable number of techniques based on precipitation in agar gels have been developed starting from the basic methods of Oudin,[41] Ouchterlony[42] and Grabar and Williams.[43] These techniques present a unique advantage over other immunochemical methods, namely the capacity of distinguishing the individual reactions due to each antigen/antibody system when a complex antigen mixture is analysed with a polyspecific immune serum. Moreover, they are easy to carry out and some of them are quite inexpensive. They are therefore also extensively used with monospecific immune sera for the qualitative and quantitative analysis of one constituent in a protein mixture. Such techniques have been widely used in food analysis (see refs. 1–8 for recent reviews). The characteristics of several of these techniques, one of which has recently been developed,[44] are compared in Fig. 2 and Table 1.

The comparisons reported in Fig. 2 (D2, E2, F2) show that ZIA is the most sensitive technique for detecting differences in antigen concentrations, followed by IED and radial diffusion. The comparisons reported in Table 1 indicate that the smallest amount of antigen is detected by electrosyneresis (1 ng/assay) and by IED and ZIA (2 and 2·5 ng/assay respectively). Electrosyneresis appears under this aspect to be 30-fold more sensitive than double diffusion according to Ouchterlony; with several

TABLE 1

COMPARISON OF SOME CHARACTERISTICS OF PRECIPITATION IN GEL TECHNIQUES SHOWN IN FIG. 1

Methods	*μl of immune serum used/assay*	*Duration (h)*			*Sensitivity*	
		Electrophoresis	*Diffusion*	*Washing*	*Amount (ng)*	*Concentration (μg/ml)*
A. Double diffusion (Ouchterlony[42])	0·5	0	48	48	30	3
B. Double diffusion (Kaminski[45])	3·6	0	48	48	10	1
C. Electrosyneresis (Bussard[46])	3·3	2	0	1	1	0·1
D. Radial diffusion (Mancini *et al.*[47])	3·3	0	168	48	4	0·4
E. IED (Laurell[49])	3·7	17	0	0·5	2	0·2
F. ZIA (Vesterberg[44])	1	24	0	0	2·5	0·05

Remarks:

1. The amount of immune serum/assay reported takes into account the dilution of the immune serum used in the experiments.
2. The duration of electrophoresis can be shortened to 30 min without changing the voltage for the electrosyneresis; it can be reduced to 6 h by increasing the voltage for IED. Low voltage and long electrophoresis duration favour the sharpness of the immunoelectrophoretic pattern for ZIA.
3. The duration of diffusion for the radial diffusion can be shortened to a few days instead of one week; however, the linearity of the relationship shown on Fig. 2 (D2) is not obtained for the higher antigen concentrations even after one week diffusion.
4. The plates have to be washed in order to remove the proteins which do not participate in the immunoprecipitation. Washing can be avoided for the following techniques: electrosyneresis, IED, ZIA.

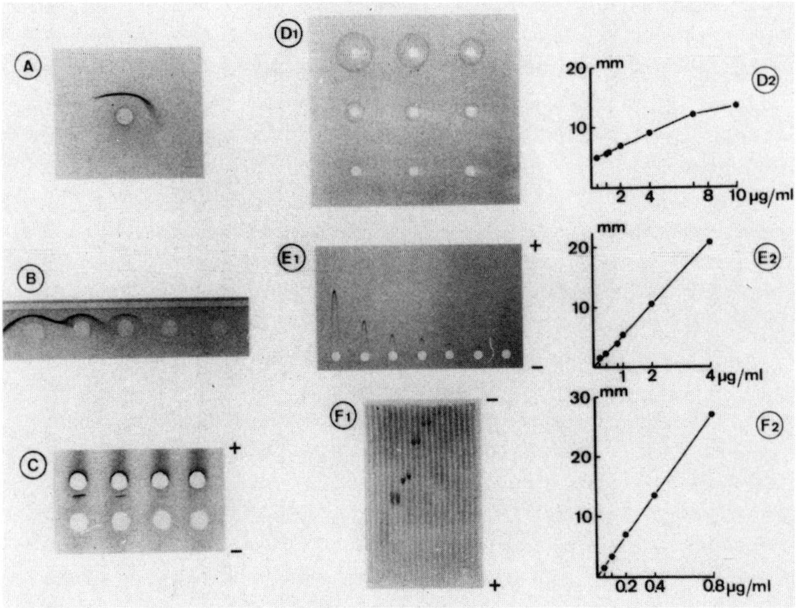

FIG. 2. Comparison between different methods of precipitation in agar gel (1·2 %) for the quantitation of one antigen by using a monospecific immune serum. In these examples, ovalbumin and an anti-ovalbumin immune serum were used for the comparison. For each technique, conditions were chosen for which a minimum amount of antigen can be detected.

(A) Double diffusion according to Ouchterlony.[42] The immune serum (diluted 3-fold here) is deposited in the central well, different dilutions of the antigen solution (10 μl) are deposited in the outer wells. Antigens and antibodies diffuse, meet, and immunecomplexes build up which form a precipitin band when a certain ratio between the amounts of antigen and antibody is reached.

(B) Double diffusion according to Kaminski.[45] The immune serum is deposited in the trough. Other conditions and remarks are as for the double diffusion according to Ouchterlony.

(C) Electrosyneresis (also called counter-immunoelectrophoresis or immuno-osmophoresis) according to Bussard.[46] The upper wells are filled with the immune serum (diluted 3-fold here) and the lower wells with different dilutions of the antigen solution (10 μl). Electrophoresis is carried out at 4°C (for 2 h here) whereby antibodies transported by the electro-endosmotic current migrate towards the cathode and the anodic antigens move in the opposite direction. They meet and form immunecomplexes and then precipitate.

(D1, D2) Single diffusion or radial immunodiffusion (RID) according to Mancini *et al.*[47] Immune serum is evenly distributed in the agar plate (0·2 % here). Wells are filled with different dilutions of the antigen solution (10 μl). During the diffusion, the antigens meet antibodies and form immunecomplexes which are first meat antigens, the detection limit of electrosyneresis was found to be 25–45-fold higher than that of double diffusion, depending on the antigen used.[51] However, as larger amounts of antigen solution can be applied in one assay for ZIA (50 μl instead of 10 μl for the other techniques in the examples shown in Fig. 2), the smallest antigen concentrations are detected by ZIA.

Two other basic techniques designed for the analysis of protein mixtures may be mentioned: immunoelectrophoretic analysis (IEA)[43] and crossed immunoelectrophoresis.[52,53] In these techniques the antigens are first separated by agarose gel electrophoresis. In a second step, for IEA, a trough is cut parallel to the migration axis and filled with the immune serum. Antigen and antibodies diffuse, meet each other and form immunecomplexes. The complexes precipitate when a certain ratio between the amounts of antigens and antibodies is reached. For crossed immunoelectrophoresis, after the first separation the agarose gel strip is embedded in an agarose gel containing the immune serum, and an electrophoresis is carried out perpendicular to the first one. The separated

---

formed in antigen excess and thus are soluble. Gradually the complexes contain more antibody and finally form a precipitin ring when optimal proportions are reached. Several parameters relating the diameter or the surface of the ring to the concentration of the antigen have been reported (see ref. 48 for review). Under the conditions used here, a linear relationship between the diameter of the ring and the concentration of the antigen is obtained (D2).

(E1, E2) Immunoelectrodiffusion (IED) (also called rocket immunoelectrophoresis) according to Laurell.[49] Immune serum (0·2 % here) is evenly distributed in the gel. Wells are filled with the antigen solutions (10 μl). Electrophoresis is carried out at a pH for which antibodies do not move or move only slightly in the gel (pH 8·6, 5 V/cm for 17 h here). Antigens migrate into the gel by electrophoresis and also by diffusion. The immunecomplexes first formed are in antigen excess but gradually the complexes collect more antibody during their migration and finally form precipitin peaks which remain unchanged during further electrophoresis. There is a linear relationship between the surface or the height of the peaks and the concentration of antigens (see ref. 50 and E2).

(F1, F2) Zone immunoassay (ZIA) according to Vesterberg.[44] The immune serum (0·2 % here) is evenly distributed in the gel which is introduced in tubes of 9 cm height and 0·2 cm diameter. The antigen solutions (50 μl) are poured into each tube above the gel surface. Electrophoresis is carried out under low voltage (50 V for 24 h). The antigens are driven in the gel by electrophoresis only and form immunecomplexes. When the immunecomplexes contain enough antibodies, they precipitate. There is a relationship between the migration distance of the immunecomplex and the antigen concentration (see ref. 44 and F2).

Remarks: In these examples, proteins in the immunecomplexes were stained with Coomassie blue after drying the gel plates.

antigens are forced to migrate in the gel containing antibodies, and they form immunecomplexes which then precipitate.

In spite of the appearance of new methods particularly well adapted for monitoring a great number of samples, such as radioimmunoassay (RIA) or enzyme-linked immunosorbent assays (ELISA), the techniques based on precipitation in gels are still widely used, probably because of their great versatility (see Tables 2 to 7). These techniques represent moreover a unique means of checking the monospecificity of an immune serum. In addition, their sensitivity can be enhanced by combining them with another technique derived from an immunohistological method. After immunoprecipitation, the plates are incubated, under standard conditions, with a preparation of goat anti-rabbit IgG antibodies conjugated with an enzyme. Subsequent assay for the activity of this enzyme increases the sensitivity of the radial immunodiffusion by a factor of 10–20.[54] Anti-rabbit IgG antibodies conjugated with a fluorescent dye may be used with electrosyneresis for the detection of soya proteins in foodstuffs.[55] Alternatively, anti-papain antibodies may be directly labelled with fluorescein isothionate; in this case the detection limit obtained by double diffusion is increased by a factor of about 5.[56]

3.3. Methods Involving the Labelling of Antigens and Antibodies

Initially, the labelling of antibodies by fluorescent labels[57] was used in histological techniques for localising an antigenic constituent in tissues or on cells. The use of ferritin,[58] enzymes[59] or protein A labelled with colloidal gold[60] extended the application of antibodies from optical microscopy to electron microscopy. The use of radioactive elements initiated considerable developments in radioimmunoassay (RIA)[61,62] for titrating one antigen or one hapten in complex mixtures. More recently, the use of enzymes as labels and the immobilisation of one of the reactants provided alternative means of solving the same problem; the method is called enzyme-linked immunosorbent assay (ELISA).[63] Both RIA and ELISA are increasingly used in food analysis; the techniques derived from immunohistology are now used only occasionally.

3.3.1. Immunohistological Methods

In immunofluorescence techniques[64] the constituents are first fixed in the tissues in order to prevent their translocation within the tissues or their leaking outside the tissue. The tissues are then incubated in the labelled antibody solution specific for the constituents to be localised. After incubation and washing, optical microscopy is applied under conditions appropriate for observing fluorescence. The fixation step must neither denature the antigenic structure of the constituent to be localised nor prevent the antibodies from reaching the antigen. (Dilutions of the immune serum or dilutions of the IgG solution are used during the incubation.) The technique is called 'direct' when the antibodies specific for the antigen are labelled, and 'indirect' when the anti-antigen antibodies are not labelled; in this case a further step is added after incubation and washing. The sample is incubated in labelled goat antibodies specific for rabbit IgG, if the first antibodies were prepared in rabbits. The labelled goat anti-rabbit IgG antibodies are commercially available. When antibodies are labelled with enzymes[59] instead of fluorescent dyes, subsequent assay for the activity of the enzyme is carried out after incubation and washing. Immunofluorescence needs expensive fluorescence microscopes. In contrast, the use of enzyme-labelled antibodies does not involve special equipment.

Although it is not widely employed in food analysis, the immunofluorescence technique is also useful for the detection of wild yeast in brewing.[65] This latter use of immunofluorescence has proved to be more sensitive than techniques based on cultures in selective media.[66] A technique using enzyme-labelled antibodies was also shown to be useful for a rapid and reliable detection of *Salmonellae* in meat and poultry products; the sensitivity was as high as that obtained with immunofluorescence.[67]

3.3.2. ELISA (Enzyme-Linked Immunosorbent Assay)

Numerous variants of ELISA have been developed (see ref. 68 for review) and several of them have begun to be used in food analysis. These techniques involve the enzyme labelling of either the antigens or the antibodies and also the immobilisation of one of the reactants on a support.[63] Peroxidase, alkaline phosphatase or glucose oxidase are used to label the antigens or antibodies. Wells in polyvinylchloride or polystyrene microtitre plates are mostly used for the immobilisation. The techniques are direct or indirect: the direct techniques involve the use of one set of antibodies usually raised in rabbits and specific for the antigen to be studied; the indirect techniques involve the use of a second set of antibodies raised in goats and specific for rabbit IgG. The second set of antibodies is labelled with the enzyme marker and is commercially available. Although the indirect techniques involve one more step in the analytical procedure, they are often found to be more convenient because no labelling of the antigen or the anti-antigen antibodies is necessary.

Each technique involves several steps, the first being the coating of the wells of the polystyrene plates with either antigens or antibodies. Each step is separated by careful washings generally with saline solutions containing low amounts of detergents in order to prevent non-specific absorptions. A brief description of the sequence of these steps is given for techniques used in food analysis. Proper blanks are carried out simultaneously, particularly assays with pre-immune serum. The enzymatic activity developing at the last step of the procedure is measured spectrophotometrically.

Variant A (an indirect technique):

1. Coating of the wells with serial amounts of the reference antigen or with the unknown samples.
2. Incubation with rabbit antibodies.
3. Incubation with enzyme-conjugated goat anti-rabbit IgG antibodies.
4. Enzymatic reaction and determination of the activity.

Remarks: The activity is in relation to the amount of antigen coated on the wells in step 1. This technique, used for detecting the source of meat in meat products, is thought to be less sensitive than the techniques involving a competition reaction.[69]

Variant B (indirect technique involving a competition reaction between bound and free antigen for the antibodies):

1. Coating of the wells with standard amounts of the antigen.
2. Incubation with standard amounts of rabbit anti-antigen antibodies mixed with serial dilutions of the reference antigen solution or with the unknown sample.
3. Incubation with enzyme-conjugated goat anti-rabbit IgG antibodies.
4. Enzymatic reaction and determination of the activity.

Remarks: The activity is inversely related to the amount of antigen present in step 2. This method was used for protein[70-72] as well as for haptenic constituents.[73-75]

Variant C (direct technique involving competition reaction between labelled and unlabelled antigens):

1. Coating of the wells with standard amounts of the rabbit anti-antigen antibodies.

2. Incubation with standard amounts of labelled antigen mixed with serial dilutions of unlabelled reference antigen solution or with the unknown sample.
3. Enzymatic reaction and determination of the activity.

Remarks: The activity is inversely related to the amount of unlabelled antigen present in step 2. The technique was used for titrating a haptenic constituent in lemon juices.[76]

Variant D (a direct technique called the double antibody sandwich ELISA technique):

1. Coating of the wells with standard amounts of unlabelled rabbit anti-antigen antibodies.
2. Incubation with serial dilutions of the reference antigen solution or with dilutions of the unknown samples.
3. Incubation with the same amounts of enzyme-conjugated rabbit anti-antigen antibodies.
4. Enzymatic reaction and determination of the activity.

Remarks: The activity is related to the amount of antigen present in step 2. The technique was used for detecting milk proteins in meat products.[77]

Variant E (an indirect technique involving the double antibody sandwich ELISA technique):

1. Coating of the wells with standard amounts of goat anti-antigen antibodies.
2. Incubation with serial dilutions of the reference antigen solution or with dilutions of the unknown samples.
3. Incubation with the same amounts of rabbit anti-antigen antibodies.
4. Incubation with the same amounts of enzyme-conjugated goat anti-rabbit IgG antibodies.
5. Enzymatic reaction and determination of the activity.

Remarks: The activity is related to the amount of antigen present in step 2. In these techniques the antibodies specific for the antigen have to be prepared in two sorts of animals, goats and rabbits. The technique was used for detecting enterotoxins in foods.[78]

The ELISA techniques represent a relatively recent methodology and improvements have to be made concerning the application of the technique to certain difficult problems. For example, for the detection of the *Clostridium botulinum* toxins, the sensitivity obtained with ELISA was found to be too low compared with the classical mouse bioassay; moreover, false positive reactions sometimes occur which are due to non-immunological bindings of proteins present in the test sample.[79] However, the ELISA techniques are among the most sensitive of existing immunochemical methods. They are often reported to be as sensitive as radioimmunoassay (RIA) techniques. In a study on enterotoxin, the sensitivity of one ELISA technique (variant E) was shown to be much higher than IED and about 10-fold higher than that reported by other authors using the passive hemagglutination technique and RIA;[78] a limit detection of 0·1 ng/ml or 0·005 ng/assay was reported. Another attractive characteristic of the ELISA technique using microtitre plates is their low immune serum consumption: the immune sera may be diluted several hundred- or thousand-fold and the volumes required for one assay range from about 20 to 100 μl. Moreover, assays on many samples can be carried out within one working day.

3.3.3. Radioimmunoassay (RIA)

General features concerning the principle of the techniques and the application of RIA to protein and hapten determinations have been reviewed.[80] The methods involve the use of radioisotopes (^{3}H, ^{14}C or ^{125}I) for labelling either the antigens[61] or the antibodies.[62] There are many RIA techniques, and they can be divided into two groups, liquid-phase RIA and solid-phase RIA.

(*a*) *Liquid-phase RIA*

The technique is often used in food analysis. Three steps can be distinguished:

1. *Incubation.* Equal amounts of the anti-antigen antibodies are incubated together or sequentially with reference unlabelled antigen or the unknown sample and with labelled antigen. The labelled antigen competes with the unlabelled one for binding with the antibodies. Since the goal of the determination is to distinguish the proportion of the labelled antigen in the free and in the antibody-bound forms, a separation has to be carried out.
2. *Separation.* The separation can be achieved by physico-chemical means when the antigen is a hapten, because of the large property

differences between the small free hapten molecule and the antibody-bound hapten. Ammonium sulphate precipitation[81,82] and charcoal absorption[83] are used in food analysis. A more specific hapten separation may be used, e.g. precipitation of the rabbit IgG using protein A.[84,85] Specific methods of separation are necessary when the antigen is a protein. In this case a second antibody set (goat anti-rabbit IgG) can be used[86] as well as protein A.

3. *Radioactivity measurement*. This is usually carried out on the pellet. In this case the radioactivity is inversely related to the amount of unlabelled antigen present in step 1.

(*b*) *Solid-phase RIA*

The technique is very similar in its principle to the ELISA technique, and can be illustrated by its application in food analysis.[87] The technique includes the following steps:

1. Coating the bottom of polystyrene tubes with the same amounts of antigen.
2. Incubation with the same amounts of rabbit immune serum and serial dilutions of the antigen or of the unknown sample.
3. Incubation with the same amounts of goat anti-rabbit IgG antibodies (antibodies are labelled with ^{125}I).

The radioactivity measured is inversely related to the amount of antigen present in step 2.

RIA techniques are used for proteins, enzymes, high molecular weight toxins as well as for low molecular compounds such as vitamins, hormones, antibiotics, pesticides and low molecular weight toxins (see ref. 9 for review and Tables 2 to 5). Numerous results already obtained with RIA for hormone determination in food products as well as methodological aspects of the determination have been reported.[9,88–90] Specific advantages of the use of ^{3}H or ^{125}I for labelling a hapten[83] and comparisons between liquid-phase and solid-phase RIA techniques[86] have been reported. RIA techniques appear well adapted to a broad field of application because of their sensitivity and the possibility of automation of the procedures.[9] However, the more recently developed ELISA techniques may be preferred to the RIA methods when both methods reach the same level of sensitivity or when the sensitivity of ELISA is sufficient for the analysis. There are several advantages for ELISA techniques: the use of stable and non-toxic reagents, the lower costs of equipment and chemicals, and absence of the problem of handling radioactive materials.

3.4. Immunoabsorption

Immunoabsorption represents a means of eliminating a given antigen from a protein mixture by using the anti-antigen antibodies. Conversely, immunoabsorption is also a means of removing certain antibodies from an immune serum by using the corresponding antigens. The latter type of immunoabsorption is mostly applied in food analysis. It aims at eliminating undesirable antibodies from an immune serum which is not monospecific in order to obtain a monospecific immune serum. A major application in food analysis is the elimination of cross-reacting antibodies from an immune serum: the object is to obtain antibodies specific for one antigen from one species only by removing the antibodies which cross-react with the homologous antigen from related species.

Immunoabsorption is traditionally carried out by using immunoprecipitation in tubes: small amounts of antigen are added to the immune serum and the precipitate is eliminated. The operation is repeated as long as a precipitate is formed.[8,91–93] Analytical immunoabsorption techniques in gels have been developed.[94–96] At present immunoaffinity chromatography[97,98] represents the best means of eliminating parasite antibodies or cross-reacting antibodies from an immune serum or from its IgG fraction. The antigens used for retaining the undesirable antibodies are immobilised on a support, Sepharose CNBr or Ultrogel for instance, and packed in a small column. The immune serum, or its IgG fraction, is passed through the column. When the column is used under non-saturated conditions, the first fraction obtained is depleted of its undesirable antibodies. The column is regenerated by desorbing the antibodies with an acidic solution and then re-equilibrated with neutral or slightly basic buffer. In these techniques the columns can be used for a great number of runs. The procedure has been used for eliminating antibodies which cross-react with bovine and ovine serum albumins from an immune serum specific for horse serum albumin. The immune serum was successively passed on two immunosorbent columns containing respectively the immobilised bovine and ovine albumins.[69] The immunoabsorbed immune serum, depleted from the cross-reacting antibodies, was then passed on a third immunosorbent column with immobilised horse albumin. The fraction obtained after desorption constituted a highly purified biological reagent containing solely anti-horse serum albumin antibodies. The antibodies were used in ELISA for controlling the origin of meat in meat products. Conversely, immunoaffinity chromatography can be used for retaining the antigen by immobilising the antibodies. An example in food analysis is given by a new technique combining immunoaffinity chromatography with RIA in order to determine staphylococcal enterotoxin B:[99] the antibodies specific for the antigen are immobilised on Sepharose CNBr columns. Reference unlabelled antigen or sample solutions are passed on to the column prior to the labelled antigen. After washing, antigens are desorbed by using a solution at pH 10·5. The radioactivity measured is inversely proportional to the amount of antigen in the reference or in the unknown sample solutions.

4. APPLICATION OF IMMUNOCHEMICAL METHODS IN FOOD ANALYSIS

There are many recent papers reviewing different aspects of immunochemical methods in food analysis.[1-9] The tables in this section, though not exhaustive, indicate present trends in this field. Two aspects are underlined: the variety of the problems investigated and the methods selected.

Tables 2 and 3 concern problems directly connected with food safety. The constituents to be detected are present in minute amounts and are often haptens, therefore very sensitive techniques have to be selected. Tables 4 to 7 mostly deal with the identification of the origin of raw materials and additives used for manufacturing food products. Much of this work concerns the adulteration of products, e.g. the use of soft wheat in flour sold for hard wheat meal, the use of cow's milk in the production of cheese sold as goat cheese, the detection of added casein, or the warranting of the animal origin of the meat used in meat proteins. Another kind of problem concerns beer, which is traditionally manufactured by using malted barley as a source of both starch and enzymes. However, part of the malt may be replaced by maize or rice as a cheaper source of starch. This procedure is accepted in many European countries but not in West Germany. Another aspect of the problem is the use of plant proteins (mainly soya proteins at present) in meat products. The use of such proteins can present technological advantages because small amounts of such proteins favour the processing of the food products. Economic reasons also favour the use of such proteins because they are cheaper than meat proteins. Therefore it is necessary to monitor the amount of such proteins in meat products. As far as additives are concerned, they may be used for technological reasons, e.g. proteolytic enzymes when added to beer prevent the formation of the chill haze. These problems were among the first to which immunochemical methods were applied in food analysis.

TABLE 2
TOXINS

Authors	*Food*	*Antigens*	*Methods*	*Limit of detection*
Lindroth and Niskanen (1977)[100]	Minced meat, sausage	Staphylococcal enterotoxin A	RIA	200 pg 2–5 ng/ml
Orth (1977)[101]	Food system	Staphylococcal enterotoxins A, B, C	RIA	1 ng/ml
Simon and Terplan (1977)[102]	Food extracts	Staphylococcal enterotoxin B	ELISA	0·1–1 ng/ml
Miller *et al.* (1978)[85]	Foods	Enterotoxins A, B, C, D, E	RIA	
Niyomvit *et al.* (1978)[99]	Non-fat milk, hamburger	Enterotoxin B	Affinity RIA	2·2 ng/ml
Robern *et al.* (1978)[103]	Food	Staphylococcal enterotoxins A, B	RIA	
Areson *et al.* (1980)[84]	Beef, buttermilk, milk, whipping cream, cheese, pastry	Enterotoxins A and B	RIA	0·1 ng/ml for A 0·5 ng/ml for B
Bergdoll and Reiser (1980)[104]	Foods	Enterotoxins A, B, C, D, E	RIA	0·4–1·3 ng/ml
Biermann and Terplan (1980)[105]	Food extracts	Aflatoxin B_1	ELISA	0·8 pg/ml
Swaminathan and Ayres (1980)[67]	Raw meat and poultry	*Salmonellae*	A direct immunoenzyme method involving microscopic observations	
Biermann and Terplan (1982)[106]	Groundnut	Aflatoxin B_1	ELISA	
Olsvik *et al.* (1982)[78]	Meat products	Enterotoxin A from *C. perfringens*	ELISA	0·1 ng/ml 1 ng/ml
Peters *et al.* (1982)[75]	Food products contaminated by fungi, e.g. mouldy corn	T2 toxin	ELISA	50 pg/ml
Morgan *et al.* (1983)[73]	Barley	Ochratoxin A	ELISA	10 pg/assay
Morgan *et al.* (1983)[74]	Potato	Solanine	ELISA	2–150 μg/100 g tissue

TABLE 3
HORMONES

Authors	*Food*	*Antigens*	*Methods*	*Limit of detection*
Hoffmann (1978)[107] Hoffmann (1981)[89] Hoffmann (1982)[108] Hoffmann and Blietz (1983)[90]	Bovine tissues	Anabolic sex hormones	RIA	ng and pg/g 10 pg for testosterone 100 pg/g for oestrogen
Allen *et al.* (1980)[83]	Cow milk	Progesterone	RIA	30 pg/assay
Vogt (1980)[109,110]	Meat, liver and kidney	Diethylstilbestrol	RIA	
Brunn *et al.* (1982)[111]	Foods products with mutton	By-product stilbene	RIA	
Günther *et al.* (1982)[112]	Foods	Phenolic oestrogens	ELISA	

TABLE 4
DRINKS

Authors	*Food*	*Antigens*	*Methods*	*Limit of detection*
Günther and Baudner (1978)[113]	Beer	Proteolytic enzymes	Double diffusion	0·001 %
			IEA	
Hebert *et al.* (1978)[114]	Beer	Proteolytic enzymes	Double diffusion	
Donhauser (1979)[91]	Beer	Bromelin, ficin, papain	Double diffusion	1 g/hl
			IEA	0·1 g/hl
			IED	
			RIA	10–25 ng
Firon *et al.* (1979)[115]	Commercial soft drinks	Orange juice	Double diffusion	5 %
Günther and Baudner (1979)[116]	Beer	Proteolytic enzymes, rice, maize	Double diffusion	
			IEA	
			Electrophoresis	
			IED	
Mansell and Weiler (1980)[76]	Lemon juice	Limonin	ELISA	
Weiler and Mansell (1980)[82]	Orange juice	Limonin	RIA	Range 0·1–100 ng
Baudner *et al.* (1982)[117]	Beer	Maize	Double diffusion	
			Electrosyneresis	
			IEA	
Fukal *et al.* (1982)[56]	Beer	Papain, chymopapain	Double diffusion	10 μg/ml
			IEA	
			Radial diffusion	5 μg/ml
			IED	2 μg/ml
			Nephelometry	2 μg/ml
Uhlig and Günther (1982)[118]	Beer	Maize, rice, papain, pepsin, bromelin, ficin	Double diffusion	
			IEA	
			Electrosyneresis	
Vaag and Gibbons (1982)[8]	Beer	Ficin, papain	Double diffusion	

TABLE 5
DAIRY PRODUCTS

Authors	*Food*	*Antigens*	*Methods*	*Limit of detection*
Harder and Chu (1979)[81]	Dairy products	Aflatoxin M_1	RIA	1–10 ng
Gombocz *et al.* (1981)[119]	Sheep milk, sheep cheese	Cow's milk, bovine casein	Double diffusion	
Lebreton *et al.* (1981)[36]	Milk	IgG (colostrum)	Immunonephelometry	
Radford *et al.* (1981)[120]	Goat's milk	Cow's milk	IED	1 %
Collin *et al.* (1982)[121]	Commercial coagulants for milk	Milk-clotting enzymes	Double diffusion	1 %
Elbertzhagen and Wenzel (1982)[122]	Sheep cheese	Cow's milk	Crossed immunoelectro-phoresis	
Wie and Hammock (1982)[123]	Whole milk	Diflubenzyron residues	ELISA	40 ppb

TABLE 6(a)
ORIGIN OF MEAT

Authors	*Food*	*Antigens*	*Methods*	*Limit of detection*
Flego and Borghese (1977)[124]	Raw and partially cooked meat products	Various kinds of meat	Electrosyneresis	
Poli *et al.* (1977)[125]	Cooked sausages	Origin of meat (bovine, porcine and equine meat)	Double diffusion	
Ponti *et al.* (1978)[126]	Meat products	Bovine, porcine and equine meat	Double diffusion IED Electrosyneresis	
Manz (1979)[92] Manz (1979)[93] Manz (1980)[127]	Mixture of meat of various animal species	Native muscle proteins	Double diffusion IEA Immunoabsorption	
Etherington and Sims (1981)[128]	Meat products	Collagen	ELISA	
Hayden (1981)[129]	Beef sausages	Equine, porcine, ovine and avian (turkey and chicken) kidney/adrenals	Diffusion	Mammalian: 5% Avian: 10%
Kang'ethe *et al.* (1982)[69]	Meat products	Horse and beef meat	ELISA	Between 3 and 8% horse in beef
Gombocz and Petuely (1983)[130]	Meat products	Bovine and swine blood plasma	Immunoturbidimetry	
Schweiger *et al.* (1982)[131]	Meat (turkey)	Muscle protein: troponin T	Double diffusion IEA Electrosyneresis	1%

TABLE 6(b)
IDENTIFICATION OF FOREIGN PROTEINS IN MEAT PRODUCTS

Authors	*Food*	*Antigens*	*Methods*	*Limit of detection*
Dougherty (1977)[132]	Sausages: frankfurters	Whey protein fraction and whey protein concentrate (α-lactalbumin and β-lactoglobulin)	IED	
Koh (1978)[133]	Mixtures of cooked and uncooked beef and soya proteins	Beef and soya proteins	Double diffusion IED	Soya: 0·14 mg/ml Beef: 0·53 mg/ml
Brehmer and Gerdes (1978)[134]	Low heated sausages	α-Casein	IED	
Poli *et al.* (1979)[55]	Heated meat products	Soya protein	Electrosyneresis with indirect immunofluorescence	Heated products 125 °C, 25 min: 2·5% with immunofluorescence
Brehmer and Gerdes (1980)[135]	Heated sausages	α-Casein	IED	0·05%
Hitchcock *et al.* (1981)[71]	Meat products	'Renatured' soya proteins	ELISA	0·1 μg/ml
Sinell and Mentz (1982)[136]	Meat products	Milk protein	IED Double diffusion precipitation ELISA	
Staak and Kämpe (1982)[70]	Sausages	Hydrolysed milk protein		
Teufel and Sacher (1982)[77]	Meat products	Milk protein	ELISA	1%

TABLE 7
MISCELLANEOUS

Authors	*Food*	*Antigens*	*Methods*	*Limit of detection*
Mazzardi (1977)[137]	Spaghetti	Soft wheat	Radial diffusion	
Klostermeyer and Offt (1978)[138]	Heated food, feedstuffs	Casein	IED	
Flego and Borghese (1979)[139]	Food products	Soya protein, whey, casein and egg	Electrosyneresis	
Gombocz *et al.* (1981)[40]	Sheep cheese, food products	Bovine casein, bovine whey protein, gliadin, soya proteins and chicken ovalbumin	Immunoturbidimetry	
Manz (1981)[140]	Egg products	Egg yolk, egg white, musculature and blood serum of the chicken	Double diffusion IEA Immunoabsorption	

Immunochemical methods are now being applied in many other fields of food analysis (Tables 2 to 7). Some of these applications were made possible because of transposition of the RIA techniques, well developed in clinical laboratories, to problems of food analysis concerning toxin and hormone determinations. The application of RIA in food analysis is relatively recent (see ref. 9 for review). However, as already mentioned, mainly because of the costs of equipment and conditions required for using radioisotopes, it is to be expected that in food analysis RIA will be less developed than ELISA. Moreover the latter technique has only recently been applied in this field.

The methods of specific precipitation in gels are still commonly used and because of their versatility they will continue to be helpful in food analysis. The techniques of immunoturbidimetry have proved to be well adapted to monitoring a large number of samples thanks to the combination of highly automated equipment. However, it needs more immune serum for the analysis than other methods. Techniques of indirect agglutination appear to be rarely used now in food analysis.

On the application of immunochemical methods in food analysis, the following comments can be made. It must first be emphasised that the control of materials contained in food may be particularly important for dietary products, such as the detection of glutenin or gliadin in dietary meals.[1] The quality of the food may be either improved or deteriorated by heat treatments: in a study of the effect of heat on soybean meal, it was suggested that an immunochemical technique, IED, could be used for uncovering the thermal pretreatment of the meal and give an evaluation of its nutritional value.[141] This was based on parallel nutritional experiments carried out with rats; it was found that the efficiency or the detrimental effect of heat treatment depends on several factors (temperature, duration of the treatment, moisture content). The methods can also prove to be helpful for evaluating materials used in different steps of food processing. For instance, immunofluorescence can be used for detecting wild yeast in the brew yeast.[65] Legislation in certain countries does not allow the addition of gibberellic acid to barley seeds to accelerate germination; an immunochemical approach seems to be specific and sensitive enough for detecting foreign gibberellic acid in malt.[142] Enzymatic activities are important markers in some food industries. However, difficulties arise when one enzyme interferes in the determination of another. That is the case for the determination of β-amylase in the presence of α-amylase because no commercial substrate is available which is specific for β-amylase only. The immunoabsorption technique which removes specifically α-amylase can provide an answer to the problem[96,143] and has been used to study the evolution of β-amylase during brewing.[144,145]

5. CONCLUDING REMARKS

Since immunochemical methods involve the use of a biological reagent, i.e. antibodies, their preparation and their properties have to be considered. First of all, there is the choice of the immunogen. Another point concerns the immunochemical determination of one antigen which may have undergone a denaturation during food processing. Also, the use of very sensitive techniques like ELISA and RIA may cause difficulties due to non-specific reactions. Finally, a major problem is the cross-reactivity of the antibodies specific for one constituent. These different points will be discussed on the basis of recent literature.

5.1. Choice of the Immunogen

The choice depends on the problem investigated: the question may concern the determination of one constituent for itself, such as a hormone. Alternatively it may deal with one constituent taken as a marker of the origin of a product, e.g. on milk protein for detecting the use of milk. In addition, the problem of determining denaturated antigen may also govern the choice of the immunogen.

When determining anabolic compounds used in animal feeding, it is important to know whether the compound is present in the milk or in the meat in its native form or in a metabolised form. The answer to this question will determine the choice of the immunogen (see ref. 90 for review). When the molecule is a hapten, the kind of binding chosen for conjugating the hapten to the protein carrier may influence one characteristic of the antibodies induced by the immunogen, namely their property of cross-reacting with molecules more or less structurally related to the hapten.[90]

Using one protein as a marker, e.g. for seed proteins, entails difficulties when a quantitative determination is made:[4,71] the protein taken as a marker represents only part of the total protein of the seed and its proportion may fluctuate from one seed sample to another, because of genetic or environmental differences. Moreover, the proportion of the marker in the protein preparation can be modified during the industrial process. Therefore, a certain approximation has to be expected when the total amount of protein of one source is estimated by the determination of one constituent present in the protein mixture.

It has long been recognised, especially in studies on the determination of foreign proteins in meat products, that it is often impossible to determine or even to detect antigens which have been denatured during heat treatment or texturisation. Native heat-stable antigens or denatured antigens or a mixture of both have been used as immunogens in order to solve the problem (see ref. 4 for review). These approaches have also been used in recent studies: to determine whey proteins used in frankfurters, α-lactalbumin was chosen preferentially to β-lactoglobulin because of its higher heat stability;[132] β-L-globulin was selected in order to determine swine blood plasma in meat products because of its high heat resistance versus antigenicity and solubility.[130] Particularly high heat-resistant kidney/adrenal antigens were chosen in order to distinguish meat from different origins in meat products: these antigens keep their antigenic reactivity after boiling or autoclaving at 120 °C for 30 min.[129] Conversely, another approach consisted of using the same denaturing conditions for extracting the immunogen and the test samples (10 M urea and 2% 2-mercaptoethanol), eliminating by dialysis the denaturing agents after extraction.[133]

5.2. Determination of Denatured Antigens

Antigens in a mixture with different ingredients and submitted to heat treatment can escape immunochemical determination not only because of antigenic modification in their structure but also because of reduced solubility. The determination of one antigen in an extract actually depends on the yield of the extraction. The yield of the extraction in turn may depend on the meat product and on the conditions used for the preparation of the food;[77] the interaction with certain liver proteins appears to be particularly important in this respect.[136] Several solutions for improving the extraction yield without altering the antigenicity of the proteins have recently been proposed: extraction by using performic acid in mild conditions,[136,138] treatment with sodium dodecyl sulphate followed by electrophoresis[146] and extraction with 7 M urea[132] were found to be efficient means for improving the extraction yield. Another approach for overcoming the difficulty of achieving an exhaustive extraction consists of using an immunoabsorption procedure.[70] In this approach, the fine suspension of the meat product was treated with a dilution of the immune serum so that the antibodies were expected to react not only with the solubilised antigen but also with the antigens remaining on the particles of the suspension. The consumption of antibodies was evaluated by determining the remaining free antibodies in the solution by means of an ELISA technique.

The determination of denatured antigens remains a difficult task, even if native stable antigens can be found as immunogens and if the extraction procedure can be improved. During denaturation, several epitopes may no longer be recognised by the antibodies, so that the antigen/antibody reaction becomes less intense, and more sensitive detection techniques have to be used.[55,71] Under these conditions a standard reference valuable for the determination of one antigen from samples which may have undergone different denaturation procedures would be difficult to find. Submitting all samples, including the reference sample, to the same conditions of heat denaturation before analysis has been proposed as a solution (see e.g. ref. 4 for review). The use of monoclonal antibodies specific for one stable epitope will probably be helpful in the future for such determinations.

5.3. Non-specific Reactions

The difficulties caused by non-specific reactions in immunochemical methods are well known and are perhaps one of the reasons why the passive haemagglutination technique applied in food analysis was subject to so much controversy (see ref. 4 for review) and why it was found unreliable for the detection of enterotoxin.[104,147] Such difficulties have been reported in very recent studies: ascorbate treatment of meat products was shown to create false positive reactions in immunoprecipitation tests,[31] non-specific reactions due to some food constituents were reported to interfere with the ELISA and solid-phase RIA techniques.[79,84,87] However, it was found that these constituents did not affect the liquid-phase RIA technique to the same extent[84] and that coating the tubes with formalised proteins in the solid-phase RIA technique eliminated non-specific interactions.[87] The technique itself can also create difficulties as shown by the use of protein A for precipitating the IgG of the immune serum in the liquid-phase RIA technique. The IgG of meat interfered with this precipitation; in this case, pretreatment of the meat extract with protein A before analysis solved the problem.[104] Thus, it remains essential to investigate the possible non-specific reactions before applying immunochemical techniques.[107]

5.4. Cross-reacting Antibodies

The reaction of some antibodies specific for one antigen with other molecules structurally related to the antigen represents one important factor in the use of immunochemical methods in food analysis. When these cross-reactivities are improperly checked, the result may be false. The problem has been exposed in detail concerning the immunochemical detection of ficin in beer:[8] several of the immune sera provided for the detection in fact cross-reacted with other protease preparations and even with barley constituents. Such cross-reactions are often found between antigens from different species belonging either to the animal or to the plant kingdom. Thus, the preparation of an immune serum requires careful examination for possible cross-reactions. The elimination of cross-reacting antibodies can be achieved by means of immunoabsorption procedures using the proper antigen in a soluble[8,91–93,119,127] or in a cross-linked[121] form or by means of immunoaffinity chromatography.[69] Another approach to prevent the formation of cross-reacting antibodies is worthy of note: antibodies specific for cow milk proteins were produced in goats in order to prevent formation of antibodies cross-reacting with goat milk proteins.[120]

Antibodies specific for a hapten may also provide cross-reactions with molecules structurally related to the hapten. Thus, it is necessary to test the reactivity of the antibodies with structurally related molecules which could be present in the food products.[73–75,81] These cross-reactions generate difficulties as in the case of the determination of aflatoxin M_1 in cow milk. Aflatoxin M_1 presents a potential hazard to human health; it is one of the major metabolites of aflatoxin B_1 present in contaminated feed ingested by cattle. Antibodies for aflatoxin B_1 do not cross-react with aflatoxin M_1. However, antibodies specific for aflatoxin M_1 cross-react with aflatoxin B_1 so that the presence of high levels of aflatoxin B_1 together with aflatoxin M_1 would affect aflatoxin M_1 determination. Solutions were provided either by separating aflatoxins M_1 and B_1 by passing them through a silica gel column or by determining separately aflatoxin B_1 with the corresponding antibodies.[81] In certain cases, the cross-reactivity may be interesting when the cross-reacting haptens present a similar toxicity as is the case for ochratoxins A and C from barley.[73]

In the near future, the production of monoclonal antibodies and the selection of those antibodies which do not possess undesirable cross-reactivity will solve the problem. However, the approach may be more difficult than expected if one takes into account that, in several cases, cross-reacting sites were reported with the use of monoclonal antibodies on molecules in which no homology was detected by using polyclonal antibodies.[148]

In conclusion, it can be said that immunochemical methods are increasingly appreciated in different aspects of food analysis. Some difficulties have still to be overcome. However, taking into account that the techniques are developing continually and that the production of monoclonal antibodies is extending considerably, it can be expected that the problems will find a solution and that immunochemical methods will be more widely applied in food analysis.

REFERENCES

1. Baudner, S. (1978). *Getreide, Mehl und Brot*, **32**, 330.
2. Llewellyn, J. W. (1979). *Int. Flavours Food Additive*, **10**, 115.
3. Scheck, K. (1980). *Fleischwirtschaft*, **60**, 406.
4. Olsman, W. J. and Hitchcock, C. (1980). In *Developments in Food Analysis Techniques–2*, King, R. D. (Ed.), Applied Science Publishers, London, p. 225.
5. Daussant, J. (1982). In *Recent Developments in Food Analysis*, Baltes, W., Czedik-Eysenberg, P. B. and Pfannhauser, W. (Eds), Proceedings of the 1st European Conference on Food Chemistry (Vienna, 1981), Verlag-Chemie, Weinheim, p. 215.
6. Káš, J., Fukal, L. and Rauch, P. (1981). *Chemické Listy/Svazek*, **75**, 963.
7. Káš, J., Fukal, L. and Rauch, P. (1982). *Prumysl Potravin*, **33**, 199.
8. Vaag, P. and Gibbons, G. C. (1982). *Brauwissenschaft*, **35**, 241.
9. Rauch, P., Fukal, L. and Káš, J. (1983). *Prumysl Potravin*. in press.
10. Sela, M., Schechter, B., Schechter, I. and Borek, F. (1967). *Cold Spring Harbor Symp. Quant. Biol.*, **32**, 537.
11. Sela, M. (1969). *Science*, **166**, 1365.
12. Atassi, M. Z. and Smith, J. A. (1978). *Immunochemistry*, **15**, 609.
13. Chua, N. H. and Blomberg, F. (1979). *J. Biol. Chem.*, **254**, 215.
14. Forsgren, A. and Sjöquist, J. (1966). *J. Immunol.*, **97**, 822.
15. Kabat, E. A. (1976). *Structural Concepts in Immunology and Immunochemistry*, 2nd Edn, Holt, Rinehart & Winston, New York.
16. Litman, G. W. and Good, R. A. (1978). *Comprehensive Immunology*, Plenum Press, New York.
17. Weir, D. M. (1978). *Handbook of Experimental Immunology*, 3rd Edn, Vol. 1: *Immunochemistry*, Blackwell, Oxford.
18. Chase, M. W. (1967). In *Methods in Immunology and Immunochemistry*, Vol. 1, Williams, C. A. and Chase, M. W. (Eds), Academic Press, New York, p. 209.
19. Harboe, N. and Ingild, A. (1973). *Scand. J. Immunol.*, **2** (Suppl. 1), 161.
20. Daussant, J. (1975). In *The Chemistry and Biochemistry of Plant Proteins*, Harborne, J. B. and Van Sumere, C. F. (Eds), Academic Press, London, p. 31.
21. Van Regenmortel, M. H. V. (1982). *Serology and Immunochemistry of Plant Viruses*, Academic Press, New York.
22. Deutsch, H. F. (1967). In *Methods in Immunology and Immunochemistry*, Vol. 1, Williams, C. A. and Chase, M. W. (Eds), Academic Press, New York, p. 315.
23. Fahey, J. L. (1967). In *Methods in Immunology and Immunochemistry*, Vol. 1, Williams, C. A. and Chase, M. W. (Eds), Academic Press, New York, p. 321.
24. Hjelm, H., Hjelm, K. and Sjöquist, J. (1972). *FEBS Lett.*, **28**, 73.
25. Fazekas de St Groth, S. and Scheidegger, D. (1980). *J. Immunol. Methods*, **35**, 1.
26. Van Vunakis, H. and Langone, J. J. (1980). *Methods in Enzymology: Immunochemical Techniques*, Vol. 70, Academic Press, New York.
27. Langone, J. J. and Van Vunakis, H. (1981). *Methods in Enzymology: Immunological Techniques*, Vols. 73, 74; (1982) Vol. 84, Academic Press, New York.

28. Stavitscki, A. B. (1977). In *Methods in Immunology and Immunochemistry*, Vol. 4, Williams, C. A. and Chase, M. W. (Eds), Academic Press, New York, p. 30.
29. Litwin, S. D. (1977). In *Methods in Immunology and Immunochemistry*, Vol. 4, Williams, C. A. and Chase, M. W. (Eds), Academic Press, New York, p. 115.
30. Kabat, E. A. and Mayer, M. M. (1961). *Experimental Immunochemistry*, 2nd Edn, C. C. Thomas, Springfield, IL.
31. Tizard, I. R., Fish, N. A. and Caoili, F. (1982), *J. Fd Protection*, **45**, 353.
32. Maurer, P. H. (1971). In *Methods in Immunology and Immunochemistry*, Vol. 3, Williams, C. A. and Chase, M. W. (Eds), Academic Press, New York, p. 1.
33. Li, I. W. and Williams, C. A. (1971). In *Methods in Immunology and Immunochemistry*, Vol. 3, Williams, C. A. and Chase, M. W. (Eds), Academic Press, New York, p. 94.
34. Leone, C. A. (1968). In *Methods in Immunology and Immunochemistry*, Vol. 2, Williams, C. A. and Chase, M. W. (Eds), Academic Press, New York, p. 174.
35. Joisel, F., Lannuzel, B., Lebreton, J. P., Boutleux, S. and Sauger, F. (1981). *Le Lait*, **61**, 568.
36. Lebreton, J. P., Joisel, F., Boutleux, S., Lannuzel, B. and Sauger, F. (1981). *Le Lait*, **61**, 465.
37. Wider, G., Hotschek, H., Findeis, I. and Bayer, P. M. (1979). *Lab. Med.*, **3**, 153.
38. Eisenwiener, H. G., Kindbeiter, J. M., Keller, M. and Gütlin, K. (1980). In *Centrifugal Analysers in Clinical Chemistry: Methods in Laboratory Medicine*, Vol. 1, Price, C. P. and Spencer, K. (Eds), Praeger, Eastbourne, p. 29.
39. Vojir, F. and Gombocz, E. (1982). *Öster. Chem. Z.*, **1**, 1.
40. Gombocz, E., Hellwig, E. and Petuely, F. (1981). *Z. Lebensm.-Unters. Forsch.*, **172**, 355.
41. Oudin, J. (1946). *C.R. Acad. Sci. Paris*, **222**, 115.
42. Ouchterlony, Ö. (1949). *Acta Pathol. Microbiol. Scand.*, **26**, 507.
43. Grabar, P. and Williams, C. A. (1953). *Biochim. Biophys. Acta*, **10**, 193.
44. Vesterberg, O. (1980). *Hoppe-Seyler's Z. Physiol. Chem.*, **361**, 617.
45. Kaminski, M. (1979). *La Pratique de l'Electrophorèse*, Masson, Paris.
46. Bussard, A. (1959). *Biochim. Biophys. Acta*, **34**, 258.
47. Mancini, G., Carbonara, A. O. and Heremans, J. F. (1965). *Immunochemistry*, **2**, 235.
48. Räsänen, J. A. (1974). *Immunochemistry*, **11**, 519.
49. Laurell, C. B. (1966). *Anal. Biochem.*, **15**, 45.
50. Clarke, H. G. N. and Freeman, T. (1967). In *Protides of the Biological Fluids*, Peeters, H. (Ed.), Elsevier, Amsterdam, p. 503.
51. Flego, R. and Borghese, R. (1979). *Boll. Chim. Un. Ital. Lab. Prov.*, **5**. 701.
52. Ressler, N. (1960). *Clin. Chim. Acta*, **5**, 795.
53. Laurell, C. B. (1965). *Anal. Biochem.*, **10**, 358.
54. Guesdon, J. L. and Avrameas, S. (1974). *Immunochemistry*, **11**, 595.
55. Poli, G., Balsari, A., Ponti, W., Cantoni, C. and Massaro, L. (1979). *J. Fd Technol.*, **14**, 483.

56. FUKAL, L., KÁŠ, J. and PALUSKA, E. (1982). *Scientific Papers of the Prague Institute of Chemical Technology*, E53, *Food*, Státní Pedagogické Nakladatelství, Prague, p. 145.
57. COONS, A. H. (1956). *Int. Rev. Cytol.*, **5**, 1.
58. SINGER, S. J. (1959). *Nature* (*London*), **183**, 1523.
59. AVRAMEAS, S. (1970). *Int. Rev. Cytol.*, **27**, 349.
60. ROTH, J., BENDAYAN, M. and ORCI, L. (1978). *J. Histochem. Cytochem.*, **26**, 1074.
61. YALOW, R. S. and BERSON, S. A. (1960). *J. Clin. Invest.*, **39**, 1157.
62. MILES, L. E. M. and HALES, C. N. (1968). *Biochem. J.*, **108**, 611.
63. ENGWALL, E. and PERLMANN, P. (1971). *Immunochemistry*, **8**, 871.
64. JOHNSON, G. D., HOLBOROW, E. J. and DORLING, J. (1978). In *Handbook of Experimental Immunology*, 3rd Edn, Vol. 1: *Immunochemistry*, Weir, D. M. (Ed.), Blackwell, Oxford, p. 151.
65. BOUIX, M. and LEVEAU, J. Y. (1978), *Bios*, **9**, 35.
66. HAIKARA, A. and ENARI, T. M. (1975). *Proc. Eur. Brew. Conv.*, 363.
67. SWAMINATHAN, B. and AYRES, J. C. (1980). *J. Fd Sci.*, **45**, 352.
68. VOLLER, A., BIDWELL, D. E. and BARTLETT, A. (1979). *The Enzyme Linked Immunosorbent Assay* (*ELISA*), Dynatech. Europe, Guernsey.
69. KANG'ETHE, E. K., JONES, S. J. and PATTERSON, R. L. S. (1982). *Meat Sci.*, **7**, 229.
70. STAAK, C. and KÄMPE, U. (1982). *Fleischwirtschaft*, **62**, 1477.
71. HITCHCOCK, C. H. S., BAILEY, F. J., CRIMES, A. A., DEAN, D. A. G. and DAVIS, P. J. (1981). *J. Sci. Fd Agric.*, **32**, 157.
72. RENNARD, S. I., BORG, R., MARTIN, G. R., FROIDART, J. M. and GEHRON ROBEY, P. (1980). *Anal. Biochem.*, **104**, 205.
73. MORGAN, M. R. A., MATTHEW, J. A., MCNERMEY, R. and CHAN, H. W. S. (1983). *Proceedings of the V International IUPAC Symposium on Mycotoxins and Phycotoxins* (Vienna, 1982), Pergamon Press, Oxford.
74. MORGAN, M. R. A., MCNERMEY, R., MATTHEW, J. A., COXON, D. T. and CHAN, H. W. S. (1983). *J. Sci. Fd Agric.*, **34**, 593.
75. PETERS, H., DIERICH, M. P. and DOSE, K. (1982). *Hoppe-Seyler's Z. Physiol. Chem.*, **363**, 1437.
76. MANSELL, R. L. and WEILER, E. W. (1980). *ACS Symp. Ser.*, **143**, 341.
77. TEUFEL, P. and SACHER, V. (1982). *Fleischwirtschaft*, **62**, 1474.
78. OLSVIK, O., GRANUM, P. E. and BERDAL, B. P. (1982). *Acta Pathol. Microbiol. Immunol. Scand. Sect. B*, **90**, 445.
79. NOTERMANS, S., HAGENAARS, A. M. and KOZAKI, J. (1982). In *Methods in Enzymology*, Vol. 84, Langone, J. J. and Van Vunakis, H. (Eds), Academic Press, New York, p. 223.
80. HUNTER, W. M. (1978). In *Handbook of Experimental Immunology*, 3rd Edn, Vol. 1: *Immunochemistry*, Weir, D. M. (Ed.), Blackwell, Oxford, p. 14.1.
81. HARDER, W. O. and CHU, F. S. (1979). *Experientia*, **35**, 1104.
82. WEILER, E. W. and MANSELL, R. L. (1980). *J. Agric. Fd Chem.*, **28**, 543.
83. ALLEN, R. M., REDSHAW, M. R. and HOLDSWORTH, R. (1980). *J. Reprod. Fertil.*, **58**, 89.
84. ARESON, P. D. W., CHARM, S. E. and WONG, B. L. (1980). *J. Fd Sci.*, **45**, 400.

85. MILLER, B. A., REISER, R. F. and BERGDOLL, M. S. (1978). *Appl. Environ. Microbiol.*, **36**, 421.
86. MENZEL, E. J. and GLATZ, F. (1981). *Z. Lebensm.-Unters. Forsch.*, **172**, 12.
87. MENZEL, E. J. and HAGEMEISTER, H. (1982). *Z. Lebensm.-Unters. Forsch.*, **175**, 211.
88. HOFFMANN, B. (1979). *J. Steroid Biochem.*, **11**, 919.
89. HOFFMANN, B. (1981). *Arch. Lebensmittelhyg.*, **32**, 57.
90. HOFFMANN, B. and BLIETZ, C. (1983). *J. Anim. Sci.*, in press.
91. DONHAUSER, S. (1979). *Brauwissenschaft*, **32**, 93, 119, 173, 211.
92. MANZ, J. (1979). *Berl. Münch. Tierärztl. Wschr.*, **92**, 6.
93. MANZ, J. (1979). *Fleischwirtschaft*, **59**, 408.
94. FEINBERG, J. G. (1957). *Int. Arch. Allergy Appl. Immunol.*, **11**, 129.
95. VAN REGENMORTEL, M. H. V. (1967). *Virology*, **31**, 467.
96. DAUSSANT, J. and SKAKOUN, A. (1974). *J. Immunol. Methods*, **4**, 127.
97. LOWE, C. R. and DEAN, P. D. G. (1974). *Affinity Chromatography*, Wiley, New York.
98. LIVINGSTON, D. M. (1974). In *Methods in Enzymology*, Vol. 34: *Affinity Techniques, Enzyme Purification*, Part B, Jokoby, W. B. and Wilchek, M. (Eds), Academic Press, New York, p. 723.
99. NIYOMVIT, N., STEVENSON, K. E. and MCFEETERS, R. F. (1978). *J. Fd Sci.*, **43**, 735.
100. LINDROTH, S. and NISKANEN, A. (1977). *Eur. J. Appl. Microbiol.*, **4**, 137.
101. ORTH, D. S. (1977). *Appl. Environ. Microbiol.*, **34**, 710.
102. SIMON, E. and TERPLAN, G. (1977). *Zbl. Vet. Med. B*, **24**, 842.
103. ROBERN, H., GLEESON, T. M. and SZABO, R. A. (1978). *Can. J. Microbiol.*, **24**, 436.
104. BERGDOLL, M. S. and REISER, R. (1980). *J. Fd Protection*, **43**, 68.
105. BIERMANN, A. and TERPLAN, G. (1980). *Arch. Lebensmittelhyg.*, **31**, 51.
106. BIERMANN, A. and TERPLAN, G. (1982). *Arch. Lebensmittelhyg.*, **33**, 17.
107. HOFFMANN, B. (1978). *J. Ass. Off. Anal. Chem.*, **61**, 1263.
108. HOFFMANN, B. (1982). *Fleischwirtschaft*, **62**, 95.
109. VOGT, K. (1980). *Berl. Münch. Tierärztl. Wschr.*, **93**, 144.
110. VOGT, K. (1980). *Arch. Lebensmittelhyg.*, **31**, 138.
111. BRUNN, H., STOJANOWIC, V., FLEMMIG, R., KLEIN, H., SHIRBINI, A. and BECHT, A. (1982). *Fleischwirtschaft*, **62**, 1009.
112. GÜNTHER, H. O., SANTARIUS, K. and JAHR, D. (1982). *Fresenius Z. Anal. Chem.*, **311**, 401.
113. GÜNTHER, H. O. and BAUDNER, S. (1978). *Lebensmittelchem. Gerichtl. Chem.*, **32**, 99.
114. HEBERT, J. P., SCRIBAN, R. and STROBBEL, B. (1978). *J. Am. Soc. Brew. Chem.*, **36**, 31.
115. FIRON, N., LIFSHITZ, A. and HOCHBERG, Y. (1979). *Lebensm.-Wiss. u. Technol.*, **12**, 143.
116. GÜNTHER, H. O. and BAUDNER, S. (1979). *Brauwissenschaft*, **32**, 200.
117. BAUDNER, S., SCHWEIGER, A. and GÜNTHER, H. O. (1982). *Z. Lebensm.-Unters. Forsch.*, **175**, 17.
118. UHLIG, R. and GÜNTHER, H. O. (1982). *Brauwelt*, **30**, 1348.

119. GOMBOCZ, E., HELLWIG, E. and PETUELY, F. (1981). *Z. Lebensm.-Unters. Forsch.*, **172**, 178.
120. RADFORD, D. V., TCHAN, Y. T. and MCPHILLIPS, J. (1981). *Aust. J. Dairy Technol.*, **36**, 144.
121. COLLIN, J. C., MUSET DE RETTA, G. and MARTIN, P. (1982). *J. Dairy Res.*, **49**, 221.
122. ELBERTZHAGEN, H. and WENZEL, E. (1982). *Z. Lebensm.-Unters. Forsch.*, **175**, 15.
123. WIE, S. I. and HAMMOCK, B. D. (1982). *J. Agric. Fd Chem.*, **30**, 949.
124. FLEGO, R. and BORGHESE, R. (1977). *Boll. Chim. Un. Ital. Lab. Prov.*, **3**, 341.
125. POLI, G., PONTI, W., BALSARI, A. and CANTONI, C. (1977). *Industrie Alimentari*, **16**, 87.
126. PONTI, W., BALSARI, A., POLI, G. and CANTONI, C. (1978). *Industrie Alimentari*, **17**, 407.
127. MANZ, J. (1980). *Fleischwirtschaft*, **60**, 763.
128. ETHERINGTON, D. J. and SIMS, T. J. (1981). *J. Sci. Fd Agric.*, **32**, 539.
129. HAYDEN, A. R. (1981). *J. Fd Sci.*, **46**, 1810.
130. GOMBOCZ, E. and PETUELY, F. (1983). *Dtsch. Lebensm. Rdsch.*, in press.
131. SCHWEIGER, A., HANNIG, K., GÜNTHER, H. O. and BAUDNER, S. (1982). In *Recent Developments in Food Analysis*, Baltes, W., Czedik-Eysenberg, P. B. and Pfannhauser, W. (Eds), Verlag Chemie, Weinheim, p. 449.
132. DOUGHERTY, T. M. (1977). *J. Fd Sci.*, **42**, 1611.
133. KOH, T. Y. (1978). *J. Inst. Can. Sci. Technol. Aliment.*, **11**, 124.
134. BREHMER, H. and GERDES, H. (1978). *Fleischwirtschaft*, **58**, 1517.
135. BREHMER, H. and GERDES, H. (1980). *Fleischwirtschaft*, **60**, 1374.
136. SINELL, H. J. and MENTZ, I. (1982). *Fleischwirtschaft*, **62**, 99.
137. MAZZARDI, E. (1977). *Boll. Soc. Ital. Farm. Osp.*, **23**, 191.
138. KLOSTERMEYER, H. and OFFT, S. (1978). *Z. Lebensm.-Unters. Forsch.*, **167**, 158.
139. FLEGO, R. and BORGHESE, R. (1979). *Boll. Chim. Un. Ital. Lab. Prov.*, **5**, 172.
140. MANZ, J. (1981). *Fleischwirtschaft*, **61**, 1580.
141. KOIE, B. and DJURTOFT, R. (1977). *Ann. Nutr. Alim.*, **31**, 183.
142. DONHAUSER, S. (1977). *Brauwissenschaft*, **30**, 325.
143. DAUSSANT, J. (1981). European Brewery Convention, Monograph VI, *EBC Symposium on the Relationship between Malt and Beer* (Helsinki, 1980), Brauwelt-Verlag Nürnberg, Darmstadt, p. 143.
144. MOLL, M., FLAYEUX, R., LIPUS, G. and MARC, A. (1981). *MBAA Tech. Quart.*, **18**, 166.
145. MOLL, M. (1982). *Brauwelt*, **122**, 754.
146. WEISSLER, N. L., MANGINO, M. E., HARPER, W. J. and RAMAN, R. (1981). *J. Fd Sci.*, **46**, 979.
147. BENNET, R. W., KEOSEYAN, S. A., TATINI, S. R., THOTA, H. and COLLINS, W. S. (1973). *Can. Inst. Fd Sci. Technol. J.*, **6**, 131.
148. LANE, D. and KOPROWSKI, H. (1982). *Nature (London)*, **296**, 200.

INDEX